PREFACE

Floods are shape-changers. They strike in different forms according to the natural forces behind them – the wind and rain, the height of the tide and sea level – and the terrain they encounter. They reveal hidden flaws; a tiny and barely perceptible slope accelerates the flow of water, a crack in a sea wall lets a torrent come gushing through. Sometimes they come as a nuisance – a few centimetres puddling outside a doorstep – but it's easy to underestimate their force. At worst, they obliterate cities or erase entire civilisations.

A flood is water where people don't want it – a deluge on our streets and in our houses – but they bring prosperity as well as ruin. The lushest grass for pasture grows after the river bursts its banks. Silt carried down the Nile was the foundation of ancient Egypt's agricultural bounty. The country's ancient name, Kemet, means 'black land' because of its fertile black soil.

The sea fills our dinner tables, while sea routes bring trade and new ideas. This is why so many great cities are built on plains by navigable rivers or by the coast; why floods are so recurrent and inescapable in our history; and why they loom so large in our imaginations. Noah's escape from catastrophe is just one of many ancient stories of a deluged Earth, the second most common myth, after creation stories. We must live by water, even if that means we sometimes die by it.

This is the story of how floods have changed our relationship with nature and with each other. The book begins in Britain in 1953, the hopeful year of Elizabeth II's coronation, when a storm in the North Sea caused the worst peacetime disaster in British history, a largely forgotten episode that still shapes our lives.

The storm hit land first at Spurn Point, on the Yorkshire coast, at 4 p.m. on 31 January 1953 – a Saturday afternoon. It then tore southwards along the coast, bringing with it darkening skies and torrential rain. Wind and tide combined to drive a wall of seawater inland along a 1,600-kilometre stretch of eastern England, smashing through concrete sea walls and sweeping away sand dunes before raging onwards across towns and fields. More than 300 people lost their lives in the UK and nearly 2,000 died in the Netherlands.

The North Sea floods of 1953 were one of the first disasters anywhere in the world attributed to a changing climate. Scientists who reported to an official inquiry noted that tides had been getting higher and storm surges off the English coast had been getting stronger. This was caused by rising sea levels which, in turn, was partly a consequence of the melting of polar ice and the shrinking of glaciers.

The inquiry's report states that 'for over 100 years there have been progressive increases both in the highest levels

THE SURGE

THE SURGE

The Race Against the Most Destructive Force in Nature

JEEVAN VASAGAR

MUDLARK

Mudlark
An imprint of HarperCollins*Publishers*
1 London Bridge Street
London SE1 9GF

www.harpercollins.co.uk

HarperCollins*Publishers*
Macken House, 39/40 Mayor Street Upper
Dublin 1, D01 C9W8, Ireland

First published by Mudlark 2026

1 3 5 7 9 10 8 6 4 2

© Jeevan Vasagar 2026

Jeevan Vasagar asserts the moral right to
be identified as the author of this work

A catalogue record of this book is
available from the British Library

HB ISBN 978-0-00-876645-0
PB ISBN 978-0-00-877559-9

Printed and bound in the UK using 100%
renewable electricity at CPI Group (UK) Ltd

CONTENTS

reached in exceptional storms and in the frequencies with which such levels occur'.[1] This was something new. Previously, countries had built flood defences based on the worst disasters of the past, but now British officials began to understand they faced a potentially escalating threat in future.

The melting of ice that these scientists referred to was not man-made. It is likely to have been caused by natural shifts in weather patterns. But the 1950s and subsequent decades saw humanity engaged in a spectacular transformation of our planetary climate: pumping out carbon dioxide and trapping more of the sun's heat, a change that is warming the air and the oceans, and melting the ice sheets of Greenland and Antarctica, creating a world of higher seas and more powerful storms.

The titanic floods of 1953 raise two profound questions for us when we face a greater risk of extreme weather than at any time in human history. The first is whether we can heed the warnings of scientists about a changing world and, secondly, whether our societies are capable of change on a scale we find hard to imagine.

Societies around the world have failed to grasp the lesson of 1953. Instead, as countries become wealthier and villagers swell the population of mega-cities, there is growing construction in flood-prone areas. Particularly in Asia, where more than half of mankind lives, the growth of settlements in flood zones has rapidly outpaced development in safer areas. River deltas are hemmed in by farmland. In Pakistan, the monsoon season that swells the Indus River and is essential for nourishing crops and replenishing water supplies now regularly brings lethal floods. Around 230 million people around the world live on land that is less than a metre above high tide. Far from reducing exposure to the hazards of a warming planet, humanity is doubling down and taking greater risks than ever before.

After 1953, Britain and the Netherlands developed the most advanced coastal engineering humanity has ever seen – giant moveable steel barriers alongside extensive sea walls of reinforced concrete. Artificial sea defences, when they are built and maintained properly, can be effective at holding back floods. These defences can save lives in combination with forecasting that tracks the path of a storm and the exercise of judgement over when to warn and evacuate people. But even the best engineering does not eliminate all risk. Rain and river floods can sweep in to cause devastation after a nation has shielded itself against the sea.

Climate scientists predict that, with increasing sea levels, the kind of extreme event seen in the North Sea in 1953 – a 'once per century' event for previous generations – will, by the middle of this century, occur at least once a year in many parts of the world. In a UN report, climate experts declare this kind of coastal storm will be more frequent 'in all scenarios', regardless of what humanity does to curb greenhouse gas emissions.[2]

The devastation of 1953 was caused by a storm surge, a combination of wind and low atmospheric pressure which drives seawater onto land. Of all the kinds of flooding – from rain, river or sea – the surge is the most devastating, moving at the greatest speed across the broadest area. It drives the greatest volume of water inland, carrying trees, rubble and other debris as it flows.

Storm surges sweep people and cars away, demolish buildings, bring down power lines and cripple underground transport systems and sewage works, leaving towns and cities uninhabitable for months. They have high death tolls and a vast economic impact; when Hurricane Katrina's storm surge overwhelmed New Orleans in 2005, the disaster cost more than $200 billion, making it the costliest hurricane in US history.[3] Built in the delta of the Mississippi River and just 160 kilometres from the sea, storm surges are a predictable natural hazard in New Orleans, but when Katrina hit, the city lacked an effective system of protection, partly because of an emphasis on shrinking government spending in the years when its defences were built.

The floods of 1953 transformed Britain. The failure to alert people living along the coast and the resulting loss of life led to the establishment of a national weather warning system, one of the first of its kind. Its initial job was to protect the east coast of England from another devastating storm surge. The threat to London also led to the creation of the massive Thames Barrier, 20 metres tall and spanning half a kilometre with steel gates that could be closed to protect the capital.

On the other side of the North Sea, the same floods caused even more terrible loss of life. In the Netherlands, the storm killed more than 1,800 people, shocking even the Dutch, who have always known that their country exists in defiance of the sea.

In response, the Dutch built the Delta Works, a vast engineering project that closed the mouths of three rivers, the Rhine, the Meuse and the Scheldt, shielding the area inland from another storm surge. A chain of barriers was built along the Dutch coast. The largest of these, the Eastern Scheldt storm surge barrier, has gates that can be shut against storm tides and reopened once the danger has passed. At nearly 3 kilometres long, this barrier is still the biggest flood protection system in the world.

The 1953 floods alerted Britain and the Netherlands to the danger they faced from the sea and the two countries responded with technological marvels. But it was a second disaster, four decades later, that demonstrated to the Dutch the limits of their remarkable engineering. In February 1995, heavy rains and melting snow swelled the Rhine, Meuse and Moselle rivers, causing a water surge that peaked in the Netherlands and put immense pressure on its dikes, the sand and clay barriers that keep floodwater out of fields and cities.

A quarter of a million people in the southern Netherlands were forced to evacuate. The dikes held, the rivers subsided and the evacuees returned to their homes, but the near-disaster brought the Dutch to a new understanding: that barriers alone could not hold back the water. Indeed, at times they actually increased the risk. The harder and higher a flood defence is built, the more terrible the consequences when it is breached. When a higher defence fails, the quantity of water it releases is greater.

After the 1995 floods, the Dutch decided to stop simply fighting the water, instead surrendering some of their hard-won land. Along rivers like the Rhine, the earthen dikes that held floodwater back were demolished and rebuilt further inland while river beds were dug deeper to give the water more room to dissipate. The Dutch called it making 'room for the river'.

It was a pragmatic move, recognising that water will always find a way around any barrier and that it was cheaper to sacrifice farmland than to risk the flooding of a major city. But it also represented a philosophical adjustment. After centuries of resisting the water, the Dutch shifted their emphasis from relying solely on defence – repelling a flood with concrete and steel – to prevention, allowing floodwaters to subside into marshes that slow down and absorb the water and wind gradually through wider and more snaking rivers. Dutch water defences still rely on engineering, but are working more with the grain of nature, rather than against it, reducing the potential for outright disaster. Instead of farmland with crops that could be ruined by flood, the Dutch created wetland that absorbed rivers when they burst their banks.

Resist the rising waters or accommodate them. It's a choice as old as human history. In the 11th century, legend has it that Canute the Great, ruler of the North Sea empire of England, Denmark and Norway, sought to demonstrate his omnipotence by ordering his fawning courtiers to place his throne on the shoreline so that he could command the tide not to touch his feet. The episode is likely to be fiction, invented at least a century after Canute died, but it lives on in the imagination. It's remembered as a story of hubris, while its deeper, Ozymandian point – that man is always part of the cosmos rather than the master of it – is often lost.

Like water dripping through cracked pipes, natural disasters highlight flaws in the system. In 2022, a monstrous monsoon turned parts of Pakistan into an inland sea, flooding around a tenth of the country and driving millions from their homes. It forced a reckoning between rich and poor nations. The calamity inspired calls for climate reparations, payable by the rich, industrialised countries that profited from burning so much carbon to poorer ones bearing the brunt of the consequences.

At United Nations climate talks in Egypt at the end of 2022, Pakistan's climate change minister Sherry Rehman led a group of developing countries in a push to create a fund that would help pay for the costs of climate disasters worldwide. The scenes in Rehman's country, which she described as a 'dystopic wasteland', helped focus delegates' minds. Nearly 200 countries agreed to set up the fund.

But there was more to this story. Climate change had intensified the monsoon rain that led to Pakistan's disaster, yet it was not the only culprit. Human intervention in nature had also played a significant part in the disaster. The super-monsoon swelled the Indus River, which flows down from the world's highest mountains to the Arabian Sea, but the river's floodplain was constricted by roads, railways and human settlement, leaving nowhere for the water to go. People's efforts to master and control the Indus, beginning under the British Raj, had amplified the tragedy.

Pakistan is not unique. The world's population now is more than three times bigger than it was in the 1950s. Cities have expanded into floodplains while farming and the creation of holiday resorts have destroyed natural flood defences, such as marshes and mangrove swamps.

The French author and philosopher Jean-Jacques Rousseau first pointed out that there is no such thing as a 'natural disaster'. Writing in the aftermath of the Lisbon earthquake of 1755, he declared that it wasn't nature that had built the thousands of multi-storey houses that collapsed. Nature played its part, of course, but it was human decisions that determined how many died and how difficult it was to recover. Unlike the Lisbon earthquake, climate change is a man-made challenge and it will be human ingenuity and imagination that shapes the outcome.

This could include engineering the ocean far beyond anything mankind has done before. A Dutch government oceanographer has proposed building giant dams to enclose the entire North Sea, creating one 500-kilometre barrier stretching from Scotland to Norway and a second 160-kilometre dam between southern England and western France, with the idea of keeping rising seas away from heavily populated coastlines.[4] It's a concept that sounds like pure science fiction, but experts agree that it's technically feasible, albeit at huge cost.

The first idea of heaven is a garden, the oldest conception of hell is a fire, and the primal disaster is a flood. Every account of climate change confronts the same paradox: we're the most advanced, accomplished civilisation ever to occupy this planet, with the most freedom, the longest lives, and technology that's far superior to every previous era. Yet we're risking all of this, even though we now fully understand the consequences of our actions. Other books have been written about the fire, the heat and what to do about CO2. This book is about H20 and what happens when the water comes. How will we respond and how will the floods change us?

As early as 1968, John Mercer, a British scientist working at Ohio State University, warned that the West Antarctic is a 'uniquely vulnerable and unstable' body of ice. A decade later, Mercer published a paper in the journal *Nature* that coupled this vulnerability with man's appetite for burning hydrocarbons, warning that if the consumption of fossil fuels continued to grow, the resulting temperature rise could cause a rapid loss of ice in West Antarctica, leading to a rise in sea levels.[5] 'Major dislocations in coastal cities and submergence of low-lying areas such as Florida and the Netherlands' might lie ahead, Mercer wrote.

The latest science, unavailable when Mercer was writing, provides us with unparalleled insight into how exactly Antarctica is being transformed by humanity and what this will mean for us and future generations. For centuries little more than a blank space at the bottom of maps, Antarctica will rewrite our future as its ice sheets are melted by a warming ocean.

Ed Carr, an American geographer who specialises in climate adaptation, was the lead author on the sixth report of the Intergovernmental Panel on Climate Change, the Geneva-based UN body that advises on the impacts of a warming world and how we can respond to it. The central message of the IPCC report is that the time for modest change has passed and that every additional year without action further narrows our options.

It's difficult for people to respond to this threat because it comes without a clear deadline. It's highly unlikely that, in ten or fifteen years, London will be underwater or New York will have been wiped off the map. The mounting danger doesn't mean it isn't possible to adapt to climate change. It's just that the choices get harder and more expensive.

Wealthy people and rich countries have more resources with which to prepare for a disaster, but Carr suggests they are also more likely to seek incremental changes, rather than accept that a way of life is changing and opt for radical trans-formation. With that reluctance to face the future comes increased economic fragility, as new commercial and residen-tial developments are built in wealthy but vulnerable cities.

Miami is one of the fastest-growing cities in the US, a place where affluent incomers are buying white-washed waterfront mansions with floor-to-ceiling windows. But while New York is currently 10 metres above sea level, Miami lies on ground that's around 2 metres above the ocean and built on porous limestone. Water flows easily through limestone so when the tide is higher than usual, saltwater gurgles up through the drains and seeps from underneath the lawns, turning expanses of bright green turf into salt marshes. As flood risk rises, it becomes harder for those on low incomes to get mortgages or to rebuild after flooding, while insurance companies are pulling out rather than waiting around to foot the bill for climate-related disasters.

The result is that those who remain on the south Florida coastline are among the wealthiest in society, clinging on to a fragile coastline – and an increasingly fragile way of life. It's hard to persuade people to abandon a dream home, especially if short-term trends suggest it could be sold for a handsome profit. But the consequences of building valuable property in cities at high risk of flooding affects more than the super-rich buying Miami condos; the last time the US housing market stumbled, the global financial system went with it.

It's easier to imagine fortifying a wealthy city against the sea – building higher sea walls and adding more pumps – than it is to accept that palm-studded boardwalks will return to swamp. Meanwhile, developers are taking on debt to put up new apartment buildings that will take 30 years to break even. 'The example of Miami screams this at me,' says Carr.[6] 'There's really not much we can do for Miami, where the water's coming from underneath, but we're still putting up new buildings.'

The question at the heart of this book is whether humans will continue to enjoy the illusion of complete mastery over nature, or whether sea, river and rain, supercharged by green-house gas emissions, will overwhelm any defence and demand a radical shift in our thinking. Mankind's power to shape nature is real, but it has limits. After 1953, the British and Dutch proclaimed dominion over the waves, creating miracles of engineering that protected people from the extremes of the sea.

Altering nature comes at a cost, though. For one thing, it destroys beauty. A concrete sea wall is usually built between the beach and the mainland, and once it's in place, it deflects waves and propels water back into the ocean with a force that scours away the sandy beach, eroding the landscape that enchants people. Building hard defences also carries a financial price. As climate change brings higher sea levels and more intense storms, those costs are rising. It has become increasingly difficult to justify the spending unless there are a significant number of lives and livelihoods at stake.

And it's not just a question of money. As every plumber knows, water always finds a way. Higher defences can always be overwhelmed by higher water, as in Pakistan where efforts to tame the Indus have contributed to more devastating floods. Put too much trust in sea walls and barriers across rivers and the consequence is that communities forget their past. Forget history and the risk is that countries neglect the need to maintain and adapt flood defences. Flooding was once woven into public memory. High-water marks are chiselled into the stone at the entrance to the church in the north Norfolk port of King's Lynn. They are written into place names too.

The -ey or -ea suffix is found across England, either denoting an island or a spot inland that is frequently threatened by water: Canvey on the Essex coast; the Norfolk village of Blakeney, home to England's largest grey seal colony; Chertsey, a mile west of the River Thames in Surrey's commuter belt. When heavy rain inundated homes in Chertsey, it came as a shock. Despite the reminder woven into its name, no one in the town believed they were at risk. Chelsea in London has a different root but a similar meaning; it is the 'chalk-port' – the place the Romans shipped in the chalk they used to make foundations for roads.

Forget history, and there's also a risk that countries lose touch with more flexible traditions of living with water. For centuries, the British have accepted the threat of flooding and adapted their lives to meet the risks. Anglo-Saxon farmers knew when to move their livestock to a hilltop refuge to escape a flooded river and then bring flocks back down to graze on the lush spring grass when the waters receded. The masters of medieval towns would make provision for clean drinking water, digging canals that channelled freshwater springs, while citizens prepared for flooding by moving their goods to higher floors.

A tradition of flood-proof construction has also been neglected. In drone videos of flooding in Tewkesbury, a market town built where the Severn and Avon rivers flow together, the 12th-century abbey – an imposing Norman church with a massive square tower – stands out like a beacon of safety, marooned but dry on a patch of higher ground while newer housing around it is underwater. In high streets across the country, the front doors of older buildings are set a few steps up from the pavement, while modern premises built next to them open directly onto the street.

Before the construction of sea defences, coastal communities were all too familiar with the destructive power of the tide. Dunwich in Suffolk, once the country's biggest port after London, was devastated in a huge storm surge on the night of

1 January 1286 and is one of a string of abandoned settlements along Britain's mutable coastline. It's hard to pinpoint when exactly our folk memories of flooding receded, but the floods of 1953 mark the modern turning point.

For the most part, the public supports moves to reduce greenhouse gas emissions, grasping the need to maintain a habitable planet for the next generation. But adapting to the *effects* of climate change, and particularly the increased risk of flooding, is another matter. The stoutness of our sea defences has made people feel immune to the effects of rises in sea levels, and the suggestion that townspeople might need to relocate away from the shoreline, or that space should be sacrificed to create wetland, is often greeted with anger and dismay. John Curtin, who worked for the UK's Environment Agency for three decades and led its response to several major floods, recalls being told by an MP: 'We didn't give up our beaches to the Germans. We're not giving them up to the Environment Agency.'[7]

One consequence of this reluctance to face reality is that rising seas are met with piecemeal adaptation. Curtin is fond of producing a series of images of a coastal pub on Canvey Island. The earliest is a sketch from the 19th century and features a man standing on a narrow coastal wall looking out to sea, with the pub nestled below it. A black and white photograph, from the early 20th century, shows a couple walking along a wider earthen rampart against the sea. In the third picture, showing the pub today, the sea is no longer visible behind the vast concrete sea wall. Like a time-lapse photograph, the sequence captures the slow transformation of the coast. 'Imagine a flood that breaks this wall or comes over it,' Curtin says. 'We would be up to our necks in water.'

Since the 1950s, the population of Britain's coastal towns and cities has more than tripled, to 8 million, with middle-class families drawn from cities by the opportunity to work remotely, joined by the ranks of pensioners downsizing to enjoy the sea air. The destitute, too, hear the call, with tenants on housing benefit filling up converted guesthouses in less prosperous towns, hoping that a fresh start by the sea will bring them better luck. Housing, roads and railways by the coastline have all expanded to serve this growth. Britain was already a country built to face the sea, with many of its great cities on coasts or in river estuaries, but there's much more life and wealth in harm's way than ever before.

Sea levels off the UK coast are expected to rise by half a metre by the end of this century, and with this rise, around a fifth of England's existing coastal defences will be vulnerable to failure.[8] By the middle of the century, the shoreline of north Norfolk and Devon will be vulnerable to a combination of wave erosion and rising seas. In subsequent decades, this will extend to large swaths of East Anglia and the south coast.

The country is sleepwalking towards a perilous future, in which the government will struggle to bear the cost of maintaining and reinforcing sea defences everywhere and people are likely to be forced to abandon their homes. Apocalyptic portrayals of a heating planet often focus on cities, with double-decker buses underwater or the Statue of Liberty half-submerged, but the reality is more likely to be that countries fray at the edges as protection is gradually withdrawn from smaller coastal towns and villages.

The UK is far from the front line of this global shift. The population of Tuvalu is already relocating, in stages, to Australia in the first climate evacuation of an entire country, while Indonesia is proposing to move its capital city away from low-lying Jakarta, now sinking faster than Venice.

Living in an era of more powerful and destructive floods will require change. It will require an end to burning fossil fuels, which continue to provide the vast majority of the energy that powers our civilisation. The coal, oil and gas we have already burned has set change in motion; we cannot escape rising seas, but we can avoid making it worse.

It will require collective action, from governments maintaining sea walls to individuals taking responsibility for making their own homes more resilient to a flood. And it will require a conversation about when and where the inevitable retreats will take place.

It is dangerous to think that these decisions can be postponed. Societies tend to respond to a crisis, rushing to rebuild and strengthen defences after a flood has hit – and sometimes making things worse as they do so. A handful of voices have consistently warned of the need to act before disaster strikes, saving lives, spending less money and, quite literally, fixing defences while the sun is shining rather than in the depths of a storm season.

There is cause for hope. Unlike the politicians and scientists of the 1950s, modern societies have the power to see into the future. The Met Office has harnessed vast computing power to model Britain's climate in years to come, while researchers probing beneath the ice of Antarctica are deepening our understanding of the ice sheets. People have begun to experiment with more flexible ways of living with water – from a café inside shipping containers on a north Ayrshire beach and a watersports company in a portable building on the Lincolnshire coast to an entire neighbourhood of floating houses in Amsterdam.

Rather than thinking of the coast as a line on a map, floods reveal it as a place of alchemy, where water and land mingle, where cliffs crumble and sand drifts, where plants and animals interact with rivers and sea, both adapting to and reshaping their environment – from the oysters that filter brackish water to the grass that holds sand dunes together. It's a place where the British have worked and traded for centuries, from herring fleets and North Sea oil to offshore windfarms.

My research for this book involved extensive interviews with experts on water management, scientists investigating the future of Antarctica, architects devising more flood-resilient homes and historians, sociologists and economists who have delved into the history of floods and the future implications of climate change, as well as bankers and insurance brokers who are considering its financial impact. I am grateful to them. Any errors are my responsibility, not theirs. Above all, it's been a privilege to interview many survivors of floods in Britain, the Netherlands, New Orleans and Asheville, North Carolina. This book is dedicated to them and to the people we have lost in floods.

CHAPTER 1

THE SURGE

By 1953, Herbert Abbs had been studying the sea for many years.[1] Sitting in his lookout tower in the pretty little Suffolk fishing town of Aldeburgh, he read the wind and the waves, the sun and the sky, and had learned to anticipate what nature was planning. And on the morning of 31 January he knew she was planning something dramatic.

Suffolk is a place where the water shapes and reshapes the earth. Unlike other sections of the English coastline edged with granite, here the boundary between sea and earth shifts constantly. On sunny days, gentle waves lap the shingle beach. But when the weather turns, they have the power to bring down cliffs.

The sea had always been part of Abbs' life. A former Royal Navy submariner, he'd grown up a little further north along this shape-shifting littoral and, for 14 years, had worked as a coastguard in Aldeburgh. And for the last seven days, he'd been watching the atmospheric pressure on his barometer. Winds flow towards an area of low pressure and the air rises where they meet, joining hands like witches in a coven. As the rising air cools, the moisture condenses, forming rain clouds and sometimes storms.

The barometer indicated that something ferocious was heading his way, something perhaps unprecedented in Abbs' lifetime. The spring tide had been exceptionally high and he knew that this, combined with the plunging atmospheric pressure, were preconditions for a devastating tempest. Looking down from his tower, Abbs watched the grey seas piling up, higher and higher, as fierce gusts drove the water against the shore and forced it upwards, towards the louring sky.

He and the other coastguards had conferred all week, but Abbs only realised on Saturday morning just how bad the storm would be. The wind and waves were going to combine to cataclysmic effect and sweep a wall of seawater inland. The coastguards had tried repeatedly to contact officials in London, but were met with silence. Most government ministers, it seemed, were away at their country houses for the weekend.

Throughout the day, the instincts of people like Abbs, who were familiar with the sea, far outpaced those of officials and emergency services, who were slower to apprehend the scale of what was unfolding. The military had a plan in case the sea walls were breached (named Operation King Canute, after the medieval monarch who tried to hold back the tide), but it would not be put into effect until much later.

The Met Office forecast that day was of gale-force winds and showers of hail or snow, but made no mention of coastal flooding. Around midday, the Met Office sent a telegram to

New Scotland Yard, warning the Metropolitan Police of an 'abnormally high tide in the River Thames in the next 24 hours'. The warning was passed on to riverside police stations across the capital. But the first casualties of the night were not in London or the east coast, but off the west coast, where rough seas forced open the loading doors of the *Princess Victoria*, a car ferry sailing from Scotland to Northern Ireland.[2]

The *Princess Victoria* was under the command of an experienced sailor, Captain James Ferguson. It was the first of a new kind of ship, a roll-on, roll-off ferry designed to allow vehicles to enter and exit more easily. After huge waves pounded open the *Princess Victoria*'s doors, water surged onto the car deck and she began listing starboard. Ferguson sent out the first distress signal just before 9 a.m. on Saturday: 'Vessel not under command.' At 1.15 p.m. that day, the last day of January 1953, with the ferry listing heavily into the water, he gave the command to abandon ship. The passengers and crew gathered by the lifeboats, with men gallantly helping women and children out first, but the sea was merciless.

The driving blizzard made it hard to see, while the wind whipped the waves into giant swells that smashed down onto the tiny lifeboats, capsizing many of them and tossing people into the freezing waters. While others scrambled into the remaining boats, the ferry capsized completely and sank into the deep. Ships answering Captain Ferguson's distress call searched for survivors all day and long into the night, their searchlights sweeping the raging sea, but just 44 out of 177 crew and passengers survived. Among the dead were the deputy prime minister of Northern Ireland, Maynard Sinclair, and the MP for North Down, Sir Walter Smiles.

The storm swept down across Britain two days after the full moon, when Earth, its satellite and the sun were in alignment. The combined gravitational force pulling on the planet's oceans created high 'spring tides'. These tides happen twice a month, with the name 'spring tide' coming from the fact that the water springs from the sea with unusual energy. There was a conjunction of natural forces: low atmospheric pressure allowed the sea level to rise; the moon and sun pulling together gave the tide greater energy; and the fierce northerly winds funnelled water down the narrow sea between Britain and northern Europe. The high winds and low pressure combined created the conditions to push the sea onto the land, a phenomenon known as a storm surge.

Storm winds, low pressure and tide united to overwhelm eastern England's coastal defences. As the storm travelled down the coast from its landfall on the Yorkshire coast, the height of the surge built, from more than 2 metres high in Lincolnshire to nearly 3 metres at King's Lynn on the Norfolk coast.

The surge swamped an oil refinery on the Isle of Grain, a marshy peninsula at the mouth of the Thames, then roared up the river into London, flooding factories and gasworks, and bringing down telephone lines. In Creekmouth, a riverside neighbourhood in east London, the floodwater was nearly a metre high. Icy water poured into hundreds of houses in

Silvertown in the East End and filled the streets in the working-class neighbourhoods around the Royal Docks.

There were four places on the English coast where casualties were heaviest. From north to south, they were the seaside resort of Mablethorpe on the Lincolnshire coast, a stretch of coastline from the town of Hunstanton to King's Lynn in Norfolk, the town of Jaywick in Essex, and Canvey Island, close to where the Thames empties into the North Sea.

Thousands of US airmen had been stationed on the east coast of England during the Second World War. The first to arrive had been given a guide to the British in which they were told that reserve should not be mistaken for haughtiness and that politeness did not indicate lack of fighting spirit. 'Sixty thousand British civilians – men, women and children – have died under bombs, and yet the morale of the British is unbreakable and high.'

In the 1950s, East Anglia's military bases were given a new purpose – confronting the Soviet Union. On the north Norfolk coast, where black-headed Brent geese escape the Arctic winter, RAF Sculthorpe hosted the airborne nuclear bombs intended to deter a Russian invasion of western Europe. They were carried on the B29 Superfortress, the giant aircraft used in the firebombing of Tokyo and the destruction of Hiroshima and Nagasaki.

The Americans, higher paid than British servicemen and with a reputation for flaunting that wealth, were not always popular. In a country that was ragged and austere after years of war, with treats like chocolate still subject to rationing, they brought a touch of swagger, with children especially delighted by American soldiers' habit of tossing bubblegum from the backs of their trucks as they passed. Young British men were less delighted with the romantic competition.

The nearest town to RAF Sculthorpe is Hunstanton, which faces west over the Wash, a shallow bay that opens onto the North Sea. In the 19th century, a local landowner developed the town as a seaside resort and a railway line was built from King's Lynn to bring day trippers to a sandy beach overlooked by cliffs striped an eye-catching red and white. The colours are natural, with a layer of the white chalk stained red by iron pigments.

There were sand dunes and salt marshes along this coastline, which provided a natural defence. Fierce waves expended some of their energy as they buffeted against dunes, while marshes soaked up torrents of floodwater. But much of the marshland had been drained for use in farming and here, as elsewhere along the coast, a shortage of housing meant many families lived in bungalows, converted railway carriages and holiday chalets clustered along the seafront just behind the line of dunes, in dangerous proximity to the sea.

RAF Sculthorpe was the biggest nuclear bomber base in western Europe, home to around 10,000 American servicemen. Many of these airmen and their families came to live in Hunstanton, where accommodation intended for summer holidays was let out cheaply over the winter. That particular Saturday was payday and, despite the cold, there was a celebratory atmosphere on the base, with airmen and their wives getting ready for parties that evening.

But as the storm raged south along the coast, there was an early indication that this was no ordinary gale; airmen at Sculthorpe saw the parked B29 bombers hoisted by the force of the wind and sent bobbing along the tarmac. They rushed out to tie the planes down.

Responsibility for maintaining sea defences was split between different authorities. In Hunstanton, the local river board maintained the sea wall along South Beach, a wide, sandy stretch, while the town council was responsible for the promenade. There was a gap between the two, around 300 metres wide, where the defence against the sea was a shingle ridge topped with a crest of grass. Much of the shingle behind this ridge had been extracted for use in making concrete during the war, despite the protest of the river board.

Freeman Kilpatrick, a US Air Force sergeant living in one of the wooden bungalows on Hunstanton's South Beach, had been getting ready to go out to the cinema with his wife when he stepped outside for a cigarette and spotted a stream of water coming into his front yard.[3] Kilpatrick dipped a finger in the trickle of water, tasted salt and grasped that the town was being flooded. He ran down the street through sleeting rain, calling out a warning to neighbours, urging them to move to higher ground. In the minutes it took the young airman to get down the street, the trickle had become a river.

He turned back to his own home, with his wife Sara, daughter Suellen and their babysitter Joyce inside, and found himself swimming against the surge of seawater. The sea poured through the gap between the promenade and the sea wall, swiftly flooding the heart of the town.

As the 7.27 p.m. train to King's Lynn left Hunstanton station, water swept across the tracks. A short distance outside town, the steam locomotive crashed into a wooden bungalow that had been ripped from the beach and carried along on the flood. The locomotive suffered minor damage to its brake pipes in the collision.

Coolly, the crew opened the doors of the passenger coaches, letting water run straight through the train to stop it filling up and tipping over. They then blocked up the hole in the train's brake pipes with rags and reversed back into the town.

Reis Leming was serving in a US Air Force squadron based at RAF Sculthorpe that specialised in maritime rescue when a USAF employee living in Hunstanton appealed for help, calling the base at 7 p.m. to say large parts of the town were now underwater and the water was rising.

He joined fellow American airmen who launched a boat from their base to rescue people trapped in their houses by floodwater. The men struggled to control the boat in high winds, navigating submerged fences, downed telephone wires and debris in the water. People trapped in their homes alerted the rescuers by signalling with flashlights, but the wind persisted in pushing the boat back. Barbed wire wrapped around the boat's propeller, before a fence post got jammed between the rudder and the propeller, leaving the boat and its crew drifting.

Airmen in a second rescue boat heard a faint voice crying for help, a sound hard to follow above the noise of the wind and sea. This crew found a 12-year-old boy, Donald Axford, on the roof of a nearly submerged caravan. Inside, his mother had died of exposure while his father floated face down in the water. The American airmen understood that people stranded on upper floors and rooftops were dying of cold in the time it was taking them to mount a rescue by boat, so they switched tactics.

Leming pulled on one of the rubber exposure suits that pilots wear to protect themselves from hypothermia if they ditch in the sea, grabbed a rubber dinghy and dragged it out into the flood, the water reaching up to his shoulders. Through darkness and a deafening gale, which splashed icy floodwater up into his face, Leming shoved the raft from house to house, straining to hear the sound of people's cries for help above the wind. The airman stopped when he saw faces at windows and pulled the survivors onto his little boat, before paddling them to safety.

Among those Leming rescued was the Quincy family, who had three children – daughters aged seven and three, and a nine-month-old baby boy. Helping all of them into the boat along with their pet dog, Leming paused to reassure the children that they were going for a 'nice boat ride' and it would be a lot of fun.

On South Beach, Freeman Kilpatrick fought his way back home and helped his wife, daughter and their babysitter climb up onto the roof, but the bungalow soon gave way beneath the force of the water and they found themselves clinging to wreckage in the flood. The shattered roof that they clung to was snagged on a cable, preventing it from being swept away.

Three factors made the onslaught of the water deadly here: the surprise, the speed and the fact that the ground behind

South Beach slopes downwards, making it difficult to escape as the floodwater rushed in and pummelled the flimsy houses.

The death toll in Hunstanton was 31, including 16 Americans, all of them servicemen and their families. Kilpatrick, his wife, daughter and their babysitter all survived. So did Leming, who kept going through the night, making three trips in all and rescuing 27 people, until his suit ripped and he collapsed from cold and exhaustion, falling unconscious into the water where he was spotted and rescued by a fellow airman.

In Aldeburgh, where Herbert Abbs had raised the alarm, hundreds of homes were flooded but no lives lost. The composer Benjamin Britten, who lived in the town, was away visiting friends on the night of the flood, but his seafront home Crag House was inundated. Britten hurried home to clean out the mud-caked cellar where he stored records; the vinyl survived but the labels were washed off. A dance at Aldeburgh's Jubilee Hall was interrupted by a police officer telling revellers to evacuate, though some of the party-goers were not quick enough to leave and had to be carried out by lifeboat crew.

The speed of the flood and the time it struck made the difference between life and death. At Lowestoft, also on the Suffolk coast, where the storm hit in late evening while people were still awake, a newly constructed sea wall slowed the floodwater. Jane Jarvis, a teenager whose mother ran a seafront guesthouse, had time to snatch up the family dog when their back door burst open and the tide swept into the hall. The guesthouse was four storeys tall and people from neighbouring houses sought refuge there, crowding into every room on the upper floors or sitting on the stairs. The *Guava*, a fishing trawler, set out from Lowestoft that Friday morning with 11 men on board and no trace of it was ever found. But at Lowestoft, no one died on land.

By midnight, the storm arrived on the north-east coast of Essex. Jaywick's low-lying fields and salt marsh, unsuitable for farming, had been developed before the Second World War as a 'plotland', a settlement where buyers were granted strips of land and built their own homes. Towns like these spread rapidly in Britain in the 1930s, driven by the urge to escape London, the rise of the car and the availability of cheap land. Seaside settlements like Jaywick had originally been marketed as holiday or weekend homes, but the destruction of housing by German bombs meant many poorer or older people began living in 'plotland' towns year round.

Jaywick was defended by a sea wall that took most of the impact of the storm surge, but the sea broke through a weak point and swept through the marshes to flood the town from behind. Audrey Frost was asleep in a bungalow with her husband Derek and toddler son Michael when she was woken by what she thought was rain. As the water began to spill through their letterbox and under the front door, they soon realised it was the sea. Outside, the water quickly reached the eaves of their bungalow. The couple began struggling to get their clothes on, then grasped there was no time to get dressed. Still naked from bed, Derek smashed a window and they swam out with their son, clambering up onto the roof. The sea rapidly engulfed Jaywick's flimsy houses and, though the Frost family all survived, 35 people lost their lives in the town.

The storm reached Canvey Island after midnight, eight hours after landfall and ten hours after the sinking of the *Princess Victoria*. Canvey is on the south-east coast of Essex, where the Thames flows out to sea. In the 17th century, a Dutch engineer had been hired by English landowners to build coastal walls around its marshy land. A few hundred Dutch migrants made their home on Canvey, escaping religious persecution in their homeland. By Edwardian times, its sandy beach and sea breezes made it a popular holiday destination for working-class Londoners escaping the smoggy capital. A number of them liked it so much they settled there, living mostly in prefabricated bungalows or holiday chalets. At the start of the 20th century, there were just 300 people living on the island; by 1953, its population had soared to nearly 13,000.

The Canvey Islanders were close-knit, many of them transplanted from the East End and still clinging to allegiances to London football clubs: West Ham, Spurs, Chelsea, Arsenal. It was a community that was poor by material standards, where children wore their older siblings' hand-me-downs, curtains froze to the windows in the chill of winter and roast chicken and fizzy drinks were a Christmas luxury. But it was rich in social ties. For the young, weekends were adventures spent catching newts and trespassing on farms, while festive occasions were marked by noisy street parties.

It was around 1am on Sunday morning when Reg Morgan, asleep in a bungalow on Canvey Island, was shaken awake by his wife Peggy, who had stretched her arm out in the darkness and realised that their bungalow was filling with icy water.[4] Towing along their five-year-old son Dennis, the couple swam out of bed, weaving around furniture that bobbed in sea-water. As they opened the front door, a wall of floodwater surged in.

Outside in the garden, Reg battled through waist-high water to hoist his wife and son onto the roof of their chicken shed. He heard his mother Phoebe, living next door, screaming out for help and swam through the swirling flood to kick her door down and help her escape. The older woman, too heavy to climb up onto the shed herself, gripped Peggy's hand in the darkness. A cold rain lashed down and the family was battered by the wind.

Unlike the Morgans, ten-year-old Ray Howard lived in a newly built two-storey house at the centre of the island.[5] Ray's father, a farmer with a smallholding of pigs and poultry, had been allotted a council home and they lived on a new estate with streetlights and paved roads. He was woken by his older sister who told him there was water gushing down the street. As floodwater filled the downstairs rooms, Ray scrambled out of the first-floor window wearing only his pyjamas and stood on the roof of the porch in the howling storm. A few minutes later, he got down into a rescue boat that had arrived alongside the house.

It was in the north of the island, where the sea wall was half a metre lower than in the south, that the sea roared through, punching down the Dutch-built sea defences and smashing through houses as dogs barked wildly and families cried out in terror. Windows were shattered and doors splintered to kindling as the water surged, carrying tables, chairs and garden sheds with it. The lack of any official warning meant that people first grasped the danger they were in when floodwater entered their homes.

The water had stripped away the Morgans' nightclothes as they swam out of their bungalow, so they were huddled naked on the chicken coop. As the cold numbed their fingers, Peggy lost her grip on her mother-in-law, who was swept away in the torrent. Then her husband Reg slipped off the chicken shed and disappeared into the darkness too. Rescuers were out in boats now, shining searchlights in the darkness to locate people trapped in their homes. She heard her neighbours shouting for help. And then human activity ebbed away. The voices stopped and the boats moved farther off.

Peggy was not aware of what had happened, but one of the search parties had entered the Morgans' home, found it empty and missed the figures clinging to the chicken shed. The rescue party marked the bungalow with a cross to show it had been searched. In the lonely hours that followed, she focused on keeping her son warm as Dennis kept repeating the words: 'Mummy, I'm cold.'

Chris Manser was asleep next to his older brother Ian in a chalet close to the northern edge of Canvey when the boys were woken by the agitated barking of their terrier Rufus.[6] Chris was one of nine children in a close-knit family that struggled for basic necessities; there was no running water and no heating, except for the gas stove. When Chris and Ian got out of bed, they found the room was filling up with water so shockingly cold that it roused them to full alertness in an instant.

Their father Edwin, a labourer, lit the gas light, letting them see that the water was now up to Chris's chest and rising. Ian decided to swim out to get help from neighbours, while the two adults and Chris gripped the younger children, battling to keep them afloat. Then Chris's father spotted another escape route, grasping – like many of his neighbours – that if there was no way out, the solution might be to perch above the flood. Edwin got onto a table to smash through the ceiling. One by one, he hoisted the children into the loft and got them to sit straddling the rafters. Their mother Anne sat on the windowsill with the two youngest children. One of the smaller boys said a prayer and then another sang a hymn and the rest of the family joined in.

The Manser family, and many others, lived between the outer sea defences and an inner sea wall, built when a Dutch engineer first reclaimed the island. This inner wall, with a concrete road running along the top of it, protected the heart of the town, but when the sea crossed the outer defences, it created a trap. With nowhere to dissipate, the water level rose rapidly in the outer housing estates.

Reg Stevens, a senior council official, knew how vulnerable those northern estates were to flooding.[7] Stevens, a handsome man with a thin moustache, was a former army major who had been awarded the Military Cross for gallantry in the Second World War. He was the council's engineer and surveyor, responsible for maintaining bridges, roads and coastal protection.

A few years earlier, Stevens had witnessed a surge tide reaching the top of the defences on the north of the island. He had urged the Essex River Authority, which handled sea defence along the county's long coastline, to raise the walls. But by 1953, this work had not been completed – partly because there were houses right by stretches of the wall, meaning land would have to be acquired from homeowners before work could be carried out.

Stevens had been out at the cinema in Southend that Saturday night with his wife and another couple. When he got home, a police sergeant knocked at his door to alert him to reports of water lapping over the earthworks to the north of the island. Stevens set out in his car to see for himself but, as he drove across town, the road became increasingly water-logged and he had to get out and walk, fighting his way through a fierce gale with thick flakes of snow.

At a street corner, the water was so deep and flowing so strongly that Stevens had to grasp a lamp post to stop himself from being swept away. Debris floated past him in the water and Stevens understood that the sea wall had been breached and that he needed to wake those who were still asleep. Hurrying back to his office, he called the fire brigade to order the sounding of air raid sirens, but the sirens' wails were not loud enough to pierce the howling wind, so he asked firefighters to blast warning rockets into the night sky.

The storm had toppled phone lines linking the island to the mainland and swamped the office where the night operator of the switchboard was working. Canvey's ambulance driver suggested a way to call for help; earlier that week, his ambulance had been fitted with a radio. Stevens used the radio to reach ambulance service headquarters, warn them of the unfolding disaster and asked them to contact the island's MP, Bernard Braine, who lived in Leigh, about 12 kilometres away on the Essex coast. The two men had been at a ceremony in the war memorial hall that Saturday afternoon, where Braine had praised the country's wartime grit.

Stevens judged that if his messages reached Braine, the MP could help muster a wider rescue effort. In the meantime, he was determined to get a better idea of what was happening in the island's northern housing estates. He headed back out, wading through water that was waist-deep in places, until he reached a vantage point at the top of the outer sea wall. The surge tide beyond the sea wall had dropped back, but floodwater now submerged the space between the outer and inner walls.

Stevens could see people trapped on the roofs of their bungalows, shouting for help, and, to his horror, a dead body. He suspected that the entire island would need to be evacuated.

Ray Howard recalls the feelings of terror and uncertainty that night. He remembers servicemen lifting him into the back of an army lorry and someone putting thicker clothes on top of his pyjamas. His parents, sister and two younger brothers were rescued alongside him, but many others clinging to rooftops slipped into the water and drowned or died of exposure waiting to be rescued. More than 300 people lost their lives in eastern England during the night of the storm and over 200 more are estimated to have drowned at sea, including those from the wreck of the *Princess Victoria* and the crew of the *Guava*. Fifty-eight of the dead were on Canvey.

There are two kinds of people who live by the coast: those for whom the sea is a respite from life in the city and those who make a living from fishing or sailing and know how dangerous it can be. Peggy Morgan was the former, someone who had moved to Canvey Island from Stepney in the East End.

Mike Brown was one of the latter.[8] He was brought up on a houseboat, moored in a creek to the north of Canvey, where he lived with his parents and elder brother. Like Herbert Abbs in Aldeburgh he was attuned to the sea's moods. On Saturday, he noticed that the morning tide was more than 2 metres lower than normal and saw the pressure drop rapidly on his barometer, a sign of an impending storm.

In the early hours of Sunday, he rowed out in a dinghy, through a housing estate that was now awash with floodwater. Brown called up to a man stranded on the roof of a bungalow, who refused to get into the boat and asked the young man to search for his wife instead. As he rowed to the other side of the bungalow, Brown recoiled in shock. A woman's body, obviously dead, was caught in a tree.

Brown, a fit man with experience on the water, managed to rescue 11 people in his dinghy, before taking a break, his hands bloodied from rowing. After changing his clothes and bandaging his hands, he went back out one more time to look for the man whose wife he had searched for. He found that he had died of exposure.

At their flooded home, Anne Manser tried to keep the children's spirits up by encouraging them to sing, but as time wore on, the family fell silent and stopped responding. Then, one of the boys, five-year-old Keith, fell from the rafters into the water. As dawn broke, Anne grasped that her two youngest boys, who were in a pram she was holding, were not asleep but dead.

Rescue arrived as the sun rose higher and the water receded. A search party came to the Mansers' chalet by canoe, its prow rapping sharply against the doors. The rescuers took the family out one by one, ferrying them to a stretch of higher ground. The Mansers walked along the top of the sea wall, leaving the island and arriving at a school on Benfleet, the nearest town on the mainland.

There, they were reunited with Ian, who had escaped by swimming to a neighbour's house. Three of the brothers had died – Keith, Gordon and Alan, who was, at two years old, the youngest victim of the floods.

Peggy Morgan's ordeal ended when one of the rescuers turned his dinghy back to have a last look at the Morgans' bungalow and his searchlight fell on the shed, illuminating the figures on its roof. Mother and child were helped into the dinghy, then taken by army lorry to Benfleet. At a shelter for flood evacuees, a doctor spotted that Dennis was suffering from severe hypothermia, which had made the muscles in his jaw clench together. He cut the five-year-old's face with a scalpel in an attempt to release the locked jaw and the pair were taken to hospital in Southend.

By the time Peggy and Dennis arrived, the hospital had run out of beds and survivors huddled on mattresses in the foyer with hot-water bottles tucked around them. Peggy was separated from her son and nurses told her the boy had been taken to the children's ward. Finally, a priest broke the news that Dennis had succumbed to hypothermia and been pronounced dead on arrival.

In the hours before sunrise, as more rescue boats arrived from neighbouring coastal towns, Reg Stevens advised the police to evacuate the entire town. Floodwaters were receding, but the island had been rendered uninhabitable. Power lines were down, roads were too waterlogged for food deliveries and the sewage system was inundated.

There was only one way to leave the island by road, along a bridge that connected Canvey to the mainland. As the winter sun gradually strengthened, hundreds walked along this route, many of them soaked to the skin and barefoot. Stevens grasped the need to move as many as possible before the tide turned, along with the threat that the single route off the island might itself be swamped by floodwater. A steady stream of buses and army lorries took people off Canvey as quickly as possible.

Some, fearful of looters, wanted to stay in their homes, but many more joined the exodus as the tide began to rise in the afternoon. The sea washed back through breaches in the flood defences and water levels in the town rose again. With the evacuation underway, Stevens asked council officials to make a map of the gaps in the island's defences, locating weak spots where there might be further collapses in the sea walls.

Bernard Braine, a former army officer and the Conservative MP for the constituency that included Canvey, arrived on the island around 7 a.m., when many were still trapped in their homes. The night before, Braine had invoked the Blitz spirit at the unveiling of a plaque commemorating Canvey's war dead. 'It seemed to me,' he said, 'that England was never so great as in adversity.' Braine observed the same wartime stoicism as he toured schools that had been converted to rescue centres. One old woman marvelled at the fact that 'a drop of water' had got her off Canvey, where Hitler had failed.

The fact that the flood happened on a Saturday night is likely to have saved lives, as workplaces and schools were closed and there were fewer people out in the open. However, this also meant there were no government officials available to give orders. The flood brought phone lines down and blocked roads, making it harder to call for support, with the result that emergency services in each town along the length of the east coast were unaware that the crisis was unfolding. Police were the earliest to respond, with commanders sending officers out in loudhailer cars to rouse people, banging on doors and blowing whistles.

The military's flood defence plan, Operation King Canute, was intended to deal with a minor breach in sea walls, which could be repaired by teams of airmen and soldiers with spades and sandbags. But it very quickly became clear that the disaster was far, far worse than anticipated – with hundreds of breaches in sea walls along the east coast. Troops from army, navy and air force bases rushed out into the night to evacuate civilians, while others worked frantically to shore up the worst of the damage to coastal defences.

Edwin Manser was a labourer, Reg Morgan a taxi driver, Ray Howard's father a farmer; all three families eked out a living. Even the sturdiest buildings were not immune, as the sea swept up debris and carried it along with enough force to punch holes in brick. But, everywhere, deaths were concentrated in places where the poorest people lived on the lowest ground and in the shabbiest housing. Some bungalows and chalets fell to pieces and their occupants drowned on the spot. In Felixstowe, prefab homes were ripped from their foundations and sent floating down water-logged streets. The cheapest houses looked as though they had been hit by an explosion as the pieces of wreckage left behind were so small and scattered.

The first rescuers were ordinary people, like Mike Brown in Canvey, who set out in their own small boats, but they were quickly joined by British and American servicemen.

That winter, there had been no prolonged spell of heavy rain or melting snow in the hill country where England's rivers rise. That meant that when the rivers in eastern England encountered the tidal surge from the North Sea, they were not swollen by rain or snowmelt, which would have intensified the disaster and spread it over a much wider area. It would have been particularly disastrous if the Fenland rivers had been in full spate, because the surrounding country is so low.

Still, the scale of the disaster was shocking. The river boards alerted their staff in the evening as it became clear how high the tide was going to be, but each one acted independently, without passing messages further down the coast.

River board engineers who scrambled over sand dunes with torches found concrete sea defences smashed to rubble. Many of the wooden bungalows and beach huts were now completely destroyed, while others had been swept from their original locations and dumped inland. From the North Sea, the flood surged up the River Nene as far as the town of Wisbech, 18 kilometres inland on the fens of Cambridgeshire.

On the north Norfolk coast, marshland that had been drained and turned into cropland or pasture was now submerged again, but the surviving salt marshes had absorbed the onslaught of the water and seemed untouched. In many cases, concrete sea walls had planted the seeds of their own destruction. When a wave strikes concrete, it is knocked back into the sea, scouring away some of the sand at the foot of a wall. Over time, this beach is scoured away completely, leaving the sea wall more exposed when a big wave strikes.

Most of the breaches in the sea walls involved a combination of different attacks from the sea, from chipping away at the front to penetrating inside the structure through cracks or crashing over the top of the wall to saturate the back. Sometimes a wall was directly smashed by the frontal assault of the waves, but more often the walls collapsed because the earth and sand inside them was washed away by water seeping into fissures.

When dawn broke on Sunday, any notion that the dark and confusion had exaggerated the destruction was quickly dispelled. It was much worse than anyone expected.

CHAPTER 2

COUNTING THE COST

Aerial photographs of the coast showed a drowned land, with residential streets, seaside holiday camps and great stretches of farmland now covered by seawater. The RAF's Operation Floodlight began on Sunday afternoon, with a reconnaissance flight sweeping north up the Lincolnshire coast. It exposed the fact that, alongside the worst death toll the country had experienced in peacetime, the scale of dislocation and destruction was vast.

As radio and newspapers spread word of the disaster, it was understood to be a national catastrophe. On the Monday after the floods hit, *The Times*, the self-assured voice of Britain's establishment, acknowledged that the wind and sea around the British isles rarely displayed such destructive power. Even so, the newspaper judged, 'the terrible size of the casualty lists does need explanation'.[1]

The reputation of American servicemen was transformed by the flood. As US air crews donated clothing and money, brought their bulldozers to repair dikes and fed evacuees in field kitchens, distrust evaporated. Young women who had been reluctant to be seen out with a US airman now flocked to their dances.

The contrast with British officials was unfavourable. While local officials, like Reg Stevens, and British army officers had been directing the rescue and recovery operations since Saturday, government ministries in London did not engage with the crisis until Monday morning, an astonishing delay by modern standards.

The death toll in Canvey, clearly the hardest-hit place, was still unknown and the exodus from the island continued, with nearly 4,000 people still trapped in their homes by flood-water. Along the shattered coastline, families forced to flee without possessions could be seen wandering the streets with bedding over their shoulders.

Initial criticism focused on the lack of a warning. The *Daily Mail* headline a few days after the floods asked 'Could Calamity Have Been Averted?'. After a previous storm surge that swept up the Thames to flood the capital in 1928, claiming 14 lives, the government had established a plan to warn London of an impending disaster, but there was no *national* warning system. This meant that while the 1953 storm had ranged along the coast from Yorkshire to the Thames estuary, each town dealt with disaster alone and with scarcely any time to evacuate, with the final act of the night devastating an island where people were asleep.

The consequences of that failure were stark, not just in the numbers of casualties but in the fact that the storm divided rich from poor. Places like King's Lynn, where there were fewer casualties, all tended to have brick-built houses that resisted the initial assault of the flood. These towns were also large enough to have police, firefighters and lifeboat crews ready to turn out in substantial numbers.

The poorer, new arrivals on the coast after the war were often those spurred to move there by wartime destruction of housing. The damage done by German bombers was still visible in those years, especially in London, Liverpool and Birmingham, where bomb craters were overgrown with weeds and many buildings lay in ruins. Winston Churchill, the war leader, was now back in office. He was 78 and in poor health after a stroke, struggling at times with decision-making and recalling names.

The country was depleted after a long war, with vast loans from the US and Canada helping shore up government finances. But Britain retained global prestige. While the struggle against the Axis powers was over, the Cold War was hot enough; there were British troops fighting in the Korean War and battling communist guerrillas in Malaya. London still commanded a colonial empire and bore the costs of maintaining overseas military bases from Singapore to Belize, as well as an occupation zone in Germany. The British Empire still appeared to be one of the world's great powers, but its politicians were vulnerable to the accusation that they had failed a basic domestic test: shielding an island country from the sea.

The response from Westminster was led by David Maxwell Fyfe, the Conservative home secretary. Maxwell Fyfe was a Scottish lawyer who had been a prosecuting counsel at the Nuremberg trials of leading Nazis, noted for his skilful cross-examination of Hermann Goering, the Luftwaffe's commander-in-chief. In private, he could be light-hearted, referring to Goering as 'fat boy' in letters home and adding with a touch of swagger that his questioning had got under Goering's skin, knocking the Nazi 'off his perch'.[2]

In public, his reputation was that of a man who achieved success through diligence rather than genius. In an era of changing attitudes, turning against the death penalty and becoming more liberal on homosexuality, Maxwell Fyfe was a defender of the status quo. Alongside the task of rescue and recovery, his challenge was to ensure that this disaster did not shake confidence in the government.

In the House of Commons on the Tuesday after the floods, Labour challenged Maxwell Fyfe over a memo, circulated in June the year before, advising local authorities of the need to conserve steel, which had delayed or halted work on coastal defences. Only work of 'exceptional urgency' would be allowed to proceed, the government memo warned. The home secretary responded by accusing his Labour critic of using tragedy as a 'stick with which to beat his political opponents'.

A further charge of neglect came in a letter to *The Times*, a week after the floods, when an engineer who had been in charge of sea defences for north Norfolk was bitterly critical of the failure to pay for adequate coastal protection.[3] Embankments along the Norfolk coast had been neglected and worn down where cattle crossed them, he wrote. Lack of funds had made recent maintenance efforts appear 'almost childish to anyone with any experience of angry seas'. His words echo through time to the present. In the 21st century, the costs of maintaining Britain's sea defences are increasing as more coastal walls are built, while more frequent incidents of heavy flooding inflict damage that needs to be repaired. In 2022, the Environment Agency, which is responsible for reducing flood risk in England, was forced on cost grounds to scale back a target of maintaining 98 per cent of its 'high consequence' defences – the ones that protected the most homes – aiming to keep around 94 per cent in peak condition instead.

In the aftermath of the 1953 floods, it soon became clear that the damage was so extensive that repair works would not be complete by the time of the next spring tides, on 14 February, when the sun and moon would align again to raise the height of high tide. The British would need to rely on nature's mercy rather than the country's tattered sea defences.

As well as funding the cost of sealing the breaches and housing those made homeless by the disaster, the government paid for the cost of repairs to homes. The Lord Mayor of London launched a public appeal for funds to help dependents of the flood victims and those whose homes had been damaged beyond repair. The fund quickly raised thousands of pounds, an indication of the strength of public dismay.

Britain was a country where people were accustomed to dominating the elements, to engineering rivers, marshland and coast since the time of the Stuart kings. Agriculture had been the force behind this change, especially in the eastern counties of England where engineers turned a wild and boggy terrain of water-loving reeds and trees into a flat expanse of wheat fields.

Tamed rivers ran straighter than ever before, while the drained land sank below water level, needing to be pumped to stay dry. When German submarines blockaded the country, these fields helped keep Britain from starvation, but the transformation had stripped away natural flood defences. When marshes had been drained and sand dunes were worn down, towns and farmland needed man-made defences, which cost far more to maintain.

Valuable agricultural land needed to be put back into service, but it would not have been surprising if a cold-eyed reality had set in about the cost of defending the entire coastline once the immediate shock of the flood was forgotten, especially small and unglamorous towns like Canvey and Jaywick. Bernard Braine, the island's MP, insisted that Canvey would 'rise again'. The question for his constituency and everywhere else that had been submerged was: how?

The first step was to assess the damage. Across much of the east coast, this was done from the air, with the RAF's reconnaissance flights photographing land that was still submerged to create a jigsaw map of the fractured defences. But in Canvey, council officials walked along the sodden perimeter of the island and recorded every point where the sea had broken through sections of sea walls.

Next was the vast challenge of sealing the gaps and draining the submerged land. Canvey's roads were still flooded so mechanical diggers could not be brought up to the sea defences. Instead, sandbags were brought to the walls by boat, with gangs of servicemen and council workmen packing the breaches by hand. At low tide, the men removed enough sandbags to create a channel that let the floodwater drain out to sea, filling this gap up again as the tide came back in. Across Canvey, council labourers dug trenches to speed the flow of water. Late on Tuesday 3 February, all the breaches on the island were sealed.

As water levels dropped, Stevens' men drained floodwater from the sewage station, dismantled the equipment inside and sent it away to be cleared of mud. The supply of clean water, gas and power was restored. The last survivor on the island, a 76-year-old woman, was found on Tuesday afternoon, but searchers continued to retrieve bodies. The police asked Bill Bishop, a greengrocer fond of a friendly word with his customers, to help identify victims.[4] Among those he recognised was a couple whose bodies were locked together in death after the woman had climbed onto her husband's back to escape the floodwater, a final act of devotion.

Some of those who had lost family members did not want to return, while others had no choice but to leave. At his primary school, Ray Howard recalls far fewer children in class on his first day back. They had been orphaned in the flood and relatives had taken them to live elsewhere in the country. While many of the families who endured the flood had already suffered the loss of loved ones in the war – two of Ray's brothers, seven-year-old Eric and five-year-old Peter, had been killed when a German V-1 rocket landed on Canvey in 1944 – this was different: an attack by a force of nature rather than a foreign enemy.

Once floodwater was pumped out of Canvey's streets and the breaches had been sealed up with sandbags, work began to raise the sea defences with steel sheet piles driven into the tops of the existing clay walls. But even when it was dry and had been made safer from the sea than before, people returned to find houses and shops wrecked and filled with mud and silt, while fields where cattle had grazed were now tainted with salt.

Many of the roads in Canvey were not paved and had become a morass of mud. To speed the drying of homes, RAF crews deployed hot-air blowers used to preheat aircraft engines in winter. Gypsum, a pink mineral that removes salt, was spread on inundated gardens while turf wriggling with earthworms was brought in to replace the soil fauna killed by saltwater. Aid packages arrived from around the country and overseas. A Canadian gift delighted children – a consignment of Pepsi rather than milk in school.

Stevens and other officials grasped there were deeper challenges to restoring life on the island. The return of commerce and finance was vital; persuading a high-street retailer to set up shop and bringing back mortgage lending would revive the town. Woolworths, a chain that sold toys and clothes alongside a pick-and-mix sweet counter, set up a mobile shop. But building societies refused to lend on an island with an uncertain future.

In the early weeks of February, around 30,000 labourers – more than half of whom were soldiers, airmen and sailors – had gone to work on the task of plugging breaches along the east coast, filling the gaps with sandbags and then piling earth against them. It was hand-to-hand combat against the sea. Hundreds of machines were deployed, with dragline

49

excavators lifting and moving shingle, while bulldozers shored up earth banks. Many roads were still flooded and bursts of rain, sleet and snow whipped across the men working out in the open, who trooped back to their accommodation exhausted and muddy.

The elements may have been unkind, but they were no longer destructive. There were no serious storms while the repairs took place, and the spring tides of Saturday 14 February did not rise high enough to bring further disaster. Instead, the temporary defences held. A few days later, Maxwell Fyfe gave a reckoning of the disaster in the Commons. 'The tidal conditions that night were far worse than anything previous experience could have led us to expect or provide against,' he said. The number of dead was now reported to be 307, while 32,000 people had been evacuated from their homes.

As he went through the statistics, the numbers revealed just how much damage had been done in a short time. Surveys suggested about 25,000 houses had been flooded and around 65,000 hectares of farmland had been inundated (an area equivalent to about 800 modern English farms). Maxwell Fyfe promised an inquiry into the causes of the disaster and the lessons to be learned.

The immediate shock was now clearly tinged with anger. At an inquest into the deaths on Canvey, the jury noted: 'We feel strongly that the worst consequences of this disaster might have been avoided if warning had been sent down the east coast.'[5] This was not just a freak event, outside the range of previous experience, as the home secretary had argued. It raised many troubling questions about the effectiveness of the British state.

It was obvious that an effective and coordinated warning system would have saved lives, but that was just one of many failures. The division of responsibility for sea defences had created weak points, such as the gap between the promenade and the sea wall at Hunstanton. The wartime neglect of the shoreline, coupled with financial constraints in peacetime, had left the entire coastline vulnerable.

This was a turning point for the country. The Second World War had been won, in part, through scientific ingenuity, but the collapse of sea defences suggested a failure of technological prowess in the face of the oldest threat to human existence. Britain was now clearly in the twilight of its imperial days, but remained a military, industrial and scientific power; a country with the resources to spring back from disaster.

On the afternoon of Saturday 31 January, as the storm harried the British coast, a meteorologist in the weather room of the Dutch national meteorological institute spotted the danger to his own country. Klaas Postma had a talent for combining observations of the weather with maps of previous storms that he held in his prodigious memory. The storm had been monitored since Friday night, as it developed south of Iceland, and into Saturday, as it sank the *Princess Victoria*, devastated forests in the north of Scotland then shifted south along the North Sea. The gap between eastern England and the Netherlands is just 300 kilometres and, as the storm travelled south, the sea grew shallow, intensifying the force of the

surge onto land. Driven by the northerly gale, the unprecedented storm was now heading directly towards the
Netherlands. Postma was determined to raise the alarm.[6]

The Netherlands is a country shaped by its relationship
with the water. The Dutch have made their home in a delta
where three great European rivers – the Rhine, the Meuse and
the Scheldt – flow out to the North Sea. For centuries, they
have endured storms and wrestled dry land from the sea.
Postma had no reason to believe his country's dikes would
fail that night, but he knew the storm's high winds could
inflict serious damage. Unlike Britain, the Netherlands had a
system for issuing national warnings. That Saturday, Postma
and his colleagues sent messages by telegram to government
officials and radio stations, warning of a storm from the
north-west – and of 'significant high water'.

The two moments of greatest vulnerability would be at
high tide on Saturday at 4 p.m. and Sunday morning's high
tide at 4 a.m. By early afternoon on Saturday, it was clear
from tidal gauges floating in the coastal water that the North
Sea was about to rise to an exceptional level – 5 metres above
mean sea level, with even higher waves likely in places. The
surge would have the greatest impact on the southern province of Zeeland, a rural area of salt marshes, forest and
'polders', low-lying fields nestling behind dikes. Here, the sea
was likely to reach 5.5 metres, meaning that it would be high
enough to reach the crest of the protective dikes.

By late afternoon, Postma upgraded the warning and radio
announcers read out an ominous message, telling listeners
that a severe storm was raging in the North Sea and there

was now a threat of 'dangerously high water'. Because it was a Saturday, government offices were unmanned. As in Britain, no one in authority coordinated a response.

Postma left the weather room that night to attend a classical concert with his wife, but he could think of little else besides the storm, which rattled the building during the performance. Once the musicians had taken their bows, Postma hurried back to work. Increasingly desperate, he contacted the radio stations and begged them to stay on air during the night, perhaps hoping this would be a way of providing a lifeline of information as the storm hit.

But, at midnight, after playing the Dutch national anthem, the broadcasters went silent.

CHAPTER 3

THE LOW COUNTRY

Six-year-old Ria Geluk was asleep in her family's farmhouse early on Sunday morning when the local doctor banged on the door.[1] As sleet pelted down and the wind roared in from the North Sea, the doctor shouted: 'The dike has broken, the water is coming!' The storm surge, whipped up by severe winds, had punched through the rampart that protected their low-lying farmland and the surrounding fields from flooding.

Two farm labourers tried to lead the cattle out of the barn, hoping the animals would find their way to safety in the open, but the unsettled herd refused to leave their warm shelter. The men gave up and turned for their cottages a short distance away, but the water was already high enough to block their path home.

The Geluk farmhouse was full that morning. The night before, a neighbouring couple had come by to play cards with Ria's parents and their three children had stayed the night with Ria and her older sister Joske.

Now Ria's mother Saar led all the children up to the first floor as water surged across the farm and into the house.

Waves of seawater rolled across the fields where Ria's father Toon had planted potatoes and wheat. The waves broke and churned violently against the farmhouse and the barn beside it. As the wind howled outside, Ria's mother told the children to play with toys and not to look out of the windows while the adults busied themselves bringing furniture up to the first floor. But it was hard to look away. The sea had come to their farm, and swept up in the waves were the shattered remnants of houses, and the bodies of people and farm animals.

Ria watched in horror as the flood carried off the farm dog Teddy, the black-and-white mongrel swept away clinging to a hay bale. In the settlements around them, the flimsiest houses had collapsed under the weight of water. A neighbour floated past, still alive, clinging to the roof of his house.

Late on Sunday, as the tide came back in and water rose to the first floor, Ria's father Toon led them further up. He realised that if they sought refuge in the attic, there was a risk that they might get trapped and drown there, so the Geluk family climbed up and out onto the roof instead. If the house collapsed, her father calculated, they should be able to cling to the roof and float, just as their neighbour had done. The family, the neighbours' children and the two workmen gathered on a wide, flat part of the farmhouse roof, blankets wrapped around them.

More than 1,800 people drowned in the Netherlands in February 1953. The catastrophic loss of life was the result not just of the terrible storm, but also – surprisingly for a country so defined by the sea – of a failure to prepare for the worst.

The Netherlands' sea defences and dikes had been damaged during the Second World War, but successive governments had also failed to prepare for a tempest of this severity.

Although the final death toll was terrible, it would have run into the tens of thousands (or more) had it not been for the boldness and quick thinking of a single person: Jaap Vogelaar, the mayor of the village of Nieuwerkerk aan den IJssel.[2] The village lay behind the Schielands High Sea Dike, a rampart holding back the waters of the Hollandse IJssel river, which branches off from the Rhine and flows west to join the Rotterdam waterway, which extends to the North Sea. In a low-lying country, this is the lowest-lying spot, more than 6.7 metres below sea level. That evening, as the sea surged east up the river and battered the dike, the mayor gathered men from the village pub, who rushed through sleeting rain to reinforce the crumbling earthen rampart with sandbags.

Freezing, soaked to the skin and bent double against the gale, the men saw the water cresting over the top of the dike and flooding into the fields beyond. Then a section of the dike collapsed and water came pouring through the gap, with a roar louder than the wind. The mayor knew immediately what was at stake. It was not just his village, but the densely populated crescent of central Holland further west, home to the cities of Amsterdam, The Hague and Rotterdam. All were threatened if the dike failed. A coastal flood in the Netherlands differs from a storm surge in other countries. The sea waters do not recede once the storm is spent, but settle over fields and houses while tides wash through each day and widen the breaches in the dikes.

In desperation, the mayor located Arie Evegroen, the captain of a grain barge *Twee Gebroeders* (Two Brothers) and ordered him 'in the name of the Queen' to steer his vessel into the widening gap. The captain was taken aback. There was a risk the boat would topple straight through the gap in the dike and plunge into the low-lying fields behind it. But he acquiesced and steered his empty barge head-on towards the breach. From the deck of the ship, the captain's mate tossed an anchor to the men on the dike and they buried it into the earth to hold the barge steady. Captain Evegroen then turned the rudder hard so the ship was sidelong onto the hole. The pressure of the water did the rest – jamming the barge into the gap. The men on shore hurried to pack more sandbags around it and secure it in place with ropes.

In a country that prided itself on planning, survival was often a matter of ingenuity like this – and sheer luck. There's a modestly sized statue of the skipper, bending over a boat, in Nieuwerkerk aan den IJssel today. The statue's called 'a penny on its side': the Dutch expression for a near miss.

Ria's family lived a tough but idyllic life on their small farm, little more than a few fields for crops and a herd of dairy cows. Their province, Zeeland, was at the south-west edge of the Netherlands and, as its name suggests, was half-sea, half-land – a clutch of islands and peninsulas sticking out into the North Sea.

The whole of the Netherlands is low-lying, but in the western half of the country, which includes Zeeland, rivers weave through a water-land of marsh and mist. The relationship between people and nature is sometimes a contest in which water has to be beaten back and made to serve human wishes. But the rivers flowing through the Low Country were a blessing too, bringing deposits of sediment that built up the land, creating higher ground where people could settle.

The waterlogged conditions were conducive to the creation of peat, soil that's rich in slowly decomposing plant matter. Dark brown or black in colour and spongy to the touch, peat is flammable and, when it dries out, can be used to heat a home or fire a forge; another gift of the water. Early settlers of Zeeland and the surrounding regions built their houses on the highest ground they could find. 'Here a wretched race is found,' the Roman scholar Pliny the Elder wrote in 77 AD, with the disdain of the civilised Mediterranean, 'inhabiting either the more elevated spots of land, or else eminences artificially constructed.' The Frisians, as these ancestral inhabitants of the land were known, pledged to defend their country with five weapons: sword and shield, spade and fork, and spear. Moving earth was just as vital as the ability to impale a Viking raider.

The inhabitants of the Dutch coast have a long history of resisting central control, declining to acknowledge the authority of feudal lords and, in particular, resisting the payment of taxes. The swampy terrain made it easier to snub authority, creating hiding places far from the reach of an aristocrat's henchmen.

The Dutch started to 'reclaim' land by the Middle Ages, digging ditches that drained peat swamps. But this disturbance of nature had an unintended consequence. When peat is immersed in water, the lack of oxygen reduces the activity of microbes that usually cause dead plants to decay, but as a swamp is drained and the peat is exposed to air, the plant matter in it begins to decay much faster. When that happens, the level of the land subsides. As the reclaimed land sank, the rivers naturally expanded and stretches of river turned into lakes.

As the lakes grew bigger, people took defensive measures, building canals to drain the excess water and dikes to hold the water back. Water was pumped out, creating low-lying fields: 'polders'. People built houses and farmed in the polders, surrounded by the high walls of a dike. The standard English word for the process of extracting dry land from water is 'reclamation', but this is slightly misleading, suggesting the land was always meant to be serviceable for human planting and habitation. The more apt Dutch word is *'landanwinning'*.

Land was conquered from the sea in the same way, with dikes built parallel to the shoreline – imitating the natural water-break of a sand dune – and then pumping out the seawater behind them. Zeeland's provincial coat of arms shows a lion emerging from waves, with the Latin motto *'Luctor et emergo'* – 'I struggle and emerge'.

More prosperous farms were mechanised, but Ria's father hitched horses to his plough. They lived on the island of Schouwen-Duiveland, a patchwork of sand dunes, creeks and polders, most of it below sea level. The island had been deliberately submerged by the German occupiers in 1944, who opened sluice gates to release water in the hope of deterring an Allied landing there. The land was drained in the months after the war and the work of replanting started. Ria recalls that while she fed the farm rabbits and generally helped or hindered as much as a cheerful toddler could, her parents worked long hours to restore the farm. *I struggle and emerge.*

Constant vigilance was needed to maintain the lattice of dike and polder, strengthening the earthworks and restoring fields when groundwater seeped up from the earth or rain accumulated on drained land. This perpetual tension became central to the nation's identity. The saying goes: 'God created the world, but the Dutch made the Netherlands.' In the art and poetry of the Netherlands, water is sometimes depicted as the 'water-wolf', a predator that steals back land from the people. It has entered everyday language too; when plucking up courage, the British use a nautical metaphor, 'sink or swim', while the Dutch say: *'pompen of verzuipen'* – 'pump or drown'.

The story of the little boy plugging a hole in a dike with his finger is famous outside the Netherlands, but it isn't actually Dutch; it was invented in the 19th century by an American writer who had never visited the country. The Dutch know that one finger will not stop a dike from giving way, but have been canny enough to put up statues of the boy for tourists' amusement.

Long before 1953, the Dutch faced devastating storms which didn't just cause vast loss of life and damage to property, but which reshaped the country's geography. On 14 December 1287, a storm surge in the North Sea coincided with a high tide and the sea battered the northern coast of the Netherlands, overwhelming sand dunes and natural clay barriers to sweep inland. As many as 50,000 are thought to have died in the Netherlands and northern Germany as a consequence of St Lucia's flood, named after the Christian saint whose feast is traditionally celebrated on 13 December.

In the contest between man and water, the sea had claimed a deadly victory. But the flooding that year had another significant consequence. It turned the Almere, a large lake in the north-west of the Netherlands, into an inlet of the North Sea that extended 100 km inland, called the 'Zuiderzee', or south sea. As a result, Amsterdam, which was then just a village by a dam on the Amstel River, gained direct access to the sea. An accident of the weather helped shape history.

On the night of 19 November 1421, the feast of St Elisabeth, another massive storm surge engulfed the south-west provinces of the Netherlands. A protracted civil war between noblemen in South Holland province, just north of Zeeland along the coast, had left the dikes there poorly maintained, resulting in catastrophe. The sea smashed a dike and submerged 300 square kilometres of reclaimed peatland that had been turned into farmland, an area called the Grote Waard. More than two dozen villages were inundated and around 2,000 lives were lost. Over time, the salt water that flooded the Grote Waard was gradually replaced with freshwater and it became a tidal wetland called the 'Biesbosch', the forest of bulrushes.

When the St Elisabeth flood was receding, according to a popular story, survivors spotted a cat balancing on a spar of timber in the water. On closer inspection, the timber proved to be a cradle with a gurgling baby inside and, according to the legend, the cat's balancing act helped keep the cradle afloat. The dike where the baby's cradle came to shore is now known as the *Kinderdijk*, the 'children's dike', while the carefree baby and the ingenious cat became the most popular image of the flood – recreated in 19th-century art, including an oil painting by the English artist John Millais – turning ruin and death into a story of hope.

As successive disasters demonstrated, the contest between water and people was never settled. While one farmer can dig a drainage ditch, a collective vigilance was required to manage the flow of a river, keep a polder dry and hold back the sea. The 16th-century dike warden Andries Vierlingh wrote a series of treatises on water management, which were rediscovered and read with enthusiasm in the 20th century. In them, he warned his countrymen that 'diking against Neptune and his consorts is a war and we must be warlike'.

The Dutch believe that the need to band together to claim dry land and face the threat of floods planted the seeds of democracy in their country. Water boards, which began in the 12th century as a way for landowners to work together to manage the dikes, then evolved into a separate strand of government with their own elections and tax-raising powers. They even had the authority to impose the death penalty – which was carried out on at least one occasion as the punishment for murdering a dike supervisor. The water boards are older than the Dutch parliament and the phrase the 'polder model' is sometimes used to describe Dutch politics, referring to the patient negotiation of an outcome from which everyone benefits. It was a kind of democracy, but not

one built on equality. Instead, the water boards were dominated by local landowners.

Across coastlines and river basins, it is ultimately only
states that have the financial capacity to provide large-scale
flood protection, but the history of water boards is a reminder
that some responsibility for flood defence will always lie
at community level. Wherever disaster strikes, neighbours
are often the first rescuers and friends the first to share food
and water.

By the early 17th century, new technology and financial
innovation would allow Dutch efforts to win land back from
water to grow increasingly ambitious. Dirck van Os arrived
in Amsterdam from Antwerp in 1587, aged 29, part of an
exodus of Protestants from the city after it was captured by
the Spanish and its inhabitants ordered to choose between
conversion to the Catholic faith or exile.

Van Os had taken part in the defence of Antwerp against
the Spanish and a portrait from the time shows a determined
young man with a wispy brown moustache, dressed in the
thick leather jerkin and armoured neck-piece of a militia
gunner. Van Os fled north to a new Dutch republic that had
just come into being after winning independence from Spain.
It was a country of thriving trade and entrepreneurship, and
a fluid moment in time when an ambitious young man could
make a fortune.

The Dutch republic's trade owed much to their mastery of
water; the expansion of agricultural land by creating polders
helped develop an export trade in cheese and butter from the
breed of cow that would become the world's most bountiful

milk producer, the black-and-white Holstein. The Dutch farmers' emphasis on breeding cattle was pragmatic in a flood-prone country. Crops are destroyed by incursions of seawater, but cattle can be herded to safety on higher ground – when they decide to cooperate, that is. Navigable rivers and access to the sea facilitated trade with England, Scandinavia and France.

Van Os, the son of a tapestry merchant, went into business with other merchants – many of whom were refugees from the Spanish, like him – and extended these trading expeditions further afield, to Russia and the Mediterranean. Alongside traditional Dutch exports like herring, these merchants added sugar, textile and metals, and they pooled their resources to venture ever further, including the most hazardous but potentially lucrative journey of all: the voyage to Asia to bring back spices. Van Os was one of the founders of the Dutch East India Company, a seafaring venture intended to win control of the eastern spice trade from the Portuguese.

Over a series of summer nights in 1602, potential investors were invited to van Os's house in central Amsterdam to take part in the world's first public share offering.[3] Other companies had raised capital from investors, but this was the first time that such investment was open to all – and shareholders were also allowed to transfer their shares, a crucial innovation that created the first market in a stock. One of the last to sign up, just before the share register closed on 31 August, was Neeltgen Cornelis, van Os's maid, who subscribed 100 guilders – a sum that must have been her life savings.

The spirit of commerce that led the Dutch to Asia turned closer to home a few years later when a group of investors led by van Os were granted the government's permission to drain the Beemster, a lake north of Amsterdam that had come into being through a combination of peat extraction and subsequent flooding. It covered a vast area, around 70 square km (the whole of modern Amsterdam is around 160 square km), and the lake was 4.8 metres below sea level at its lowest point. Under the direction of Jan Leeghwater, an architect and engineer, a giant dike was built in a ring around the lake, with a canal surrounding it for drainage. An array of windmills, each one pumping the water a little higher to the next one along, was then used to turn the lake into dry land, in 1612.

Van Os and his fellow traders put up the finance, while Leeghwater supplied the technical ingenuity to create farmland that proved exceptionally fertile. Milk from cows reared on Beemster grass produced a smooth-textured cheese with a sharp tang. As if to underline the bringing of man-made harmony to nature, the Beemster's fields are divided into a neat geometric pattern of squares bisected by canals and roads that meet at right angles, like a giant chessboard. Dirck van Os's son, also named Dirck, would serve as the *dijkgraaf*, the head of the water board responsible for the Beemster polder. Rembrandt painted the younger Dirck's portrait as an old man, his father's determination still evident in his gaze and a ceremonial cane in his hand.

The reclamation of land transformed the Netherlands physically, creating new pastures and cropland, and shortening the coastline; where the coast once wiggled along the meandering path of sandbanks and estuaries, dikes created straight lines. The land behind the dikes was less liable to flooding than before and could be exploited by farmers and taxed by government. So wresting land from water changed Dutch society too, turning a nation whose ancestors had lived in swampland, scrappy and rebellious, into wealthy farmers and merchant princes.

The pioneers of reclamation understood themselves to be working with nature rather than against it, despite the fact that they were redirecting the flow of rivers and reshaping coastlines. Vierlingh, the 16th-century dike warden, counselled patience and working at the pace dictated by nature and the tides, taking time to make subtle changes at the right moment rather than imposing sudden and forceful alterations.

Dutch engineers exported their craft around Europe, from Italy to Russia. In Britain, a swaggering young Dutchman called Cornelius Vermuyden was hired to drain East Anglia's fens, transforming a landscape of marsh and forest into England's most fertile farmland. Met with fierce resistance from the people who traditionally made a living gathering reeds and hunting for wildfowl in the fens, Vermuyden left a lasting imprint on the English landscape, creating a flat and treeless expanse of fields below sea level, that have to be constantly pumped out to stop them being reconquered by water. Before they begin, such titanic projects have a touch of the fantastical. But the science fiction of the past becomes the ground beneath people's feet, so commonplace that it escapes notice – until something goes wrong.

When the Netherlands was under French rule in the late 18th century, a central government agency was created to manage roads and waterways; this became known as the Rijkswaterstaat, which still builds Dutch roads and manages canals and dikes today. But the water boards were reluctant to surrender control and the Rijkswaterstaat had to work alongside them.

As the Dutch established a colonial empire abroad, money flowed back to the Netherlands to fill state coffers, allowing the Rijkswaterstaat to hire more officials and grow more powerful. But by the 1950s, there were still around 2,700 water boards in the Netherlands, like the one van Os's son headed. Each was responsible for the upkeep of their own dikes and pumping out rainwater from the fields, and the water boards administered most of the country's sea walls.

When the floods struck in 1953, it became clear there were deep flaws in the Dutch model for holding back the sea. The meteorologist Klaas Postma's warning of dangerously high water went largely unheard, as only three of the water boards had a subscription to the warning service. A number of government officials had subscribed, however, including the mayors of two towns in the southern Netherlands. And when Postma returned from the concert, desperate to find someone who would listen to a warning, he got a call from one of them.

Cor van der Hooft, the mayor of Willemstad, was convinced this was no ordinary storm.[4] His phone call with Postma late that Saturday strengthened his conviction that he needed to raise the alarm. While the meteorologist anticipated that the storm winds would cause damage, likely ripping roofs off houses and snapping branches off trees, he had no reason to think the dikes would be overwhelmed.

But the Netherlands' sea defences were in a dangerously weakened state. In wartime, the German occupiers had tunnelled into the dikes, building concrete bunkers directly into the earth as they prepared for Allied attack. As the beaches and dunes along the entire coast were closed to civilian authorities during the war, pest control lapsed too and digging by rabbits and moles added to the Wehrmacht's damage. After the war, as a shattered nation began to rebuild, many of the weak spots were repaired by simply filling in the holes rather than properly assessing the damage to the structure.

When the storm hit, the weaknesses introduced by German tunnelling proved disastrous. The sea overwhelmed the dikes, which were often faced with stone on the marine side but just clay on the landward side. As waves splashed over the crest of the dikes, seawater penetrated weak spots on the landward side and saturated the clay, bringing down whole sections and letting more water rush through the gap. In all, there were nearly 150 breaches in the sea defences.

As in Britain, the response to the disaster was confused. There was no plan to guide local officials and no central coordination. But the mayors of Willemstad and Ooltgensplaat took the telegram warnings from the weather service seriously.

Ooltgensplaat's mayor Pieter Hordijk had trained as an infantry officer before the German invasion. During the Nazi occupation of his country, he had become a commander in a Dutch resistance unit known as the LKP, the 'thug squad', specialising in sabotage. As the storm swept in, Hordijk was decisive, sending police and firefighters out to warn villagers and ordering them to secure homes and streets with flood barriers made out of wooden boards reinforced with sandbags.[5]

Willemstad's mayor van der Hooft appealed for help from a provincial official, who ignored him and went back to sleep. Instead, van der Hooft sent police out to wake the townspeople. When the officers came back and reported that the citizens would not leave, he ordered them to throw rocks through windows and shout a warning to escape before the flood came. Then the mayor headed to the dikes to reinforce them.

In some cases, a degree of complacency was rooted in past experience. The Dutch had witnessed and survived severe winter storms before, including one in September 1944 with hurricane-force winds. The speed at which the water advanced also made it hard to react. Evacuating in the face of the flood carried its own risks. As the wind roared and water rushed in, it was instinctive to find shelter and hide rather than dash for higher ground. But, in 1953, the chaotic official reaction would have deadly consequences.

In the village of Oude Tonge, in South Holland province, the mayor had gone to bed early, sleeping off the after-effects of a wedding party on Friday night. The village is on an island and has a small harbour linked by a canal to a tidal river that drains into the North Sea. Two of the villagers, a boatman and his brother-in-law, Aren Kanters, a council official, were unable to sleep and walked down to the canal's floodgates in the middle of the night.[6] Here, in the very early hours of Sunday morning, they found the water was just a metre below the crest of the dike that protected the village and rising rapidly. Kanters rushed back to the village, gathering the local police commander, and went to rouse the mayor around 3 a.m.

The village was still dry and the mayor hesitated, worried about giving a false alarm. His hesitation squandered the time left to save lives; by now, the sea-dike to the west of the village had broken and the polders between the dike and the village were rapidly filling with salt water. By 5 a.m., the inner dikes protecting the village began to give way and then, in a matter of minutes, a wave of seawater surged through Oude Tonge.

A cabin boy named Cor van de Tonnekreek, woken by the noise of the storm, came out to find a row of neighbours' houses already up to their eaves in water, with families scrambling onto their roofs for safety. With the help of other villagers, Cor dragged a flagpole up to the window of his house and laid it across to the gutters of the house opposite. Young and athletic, he wriggled across the horizontal flagpole with a rope tied around his waist, then secured the rope in place so that people could inch across to safety while clinging onto the pole and rope. Cor's heroic feat saved 47 people.

But, for dozens more, the mayor's hesitation was fatal. Survivors described waves smashing through the village and pounding against the walls of timber houses. Some were sleeping on the ground floor, with no time to get away as the water rushed in. Others were able to climb up onto roofs, only for their houses to collapse under the force of the water.

A total of 300 people died in Oude Tonge that day. In Ooltgensplaat, there were just two casualties, a pair of farm-workers who were trying to move a combine harvester out of the way of the flood. The men drowned in the barn where the machine was stored.

One of the first indications of the extent of the crisis came from Cor van der Hooft in Willemstad, who contacted the national press agency ANP in the early hours of Sunday. Polders in Willemstad were being flooded, he said. The agency relayed a stark telegram to newsrooms across the country: 'Military assistance has been called in. Power is out. The town is being submerged.' Many telephone lines were down, but the reports that got through made clear that this was a flood unparalleled in living memory. Dutch radio broadcast the first sombre bulletin that Sunday morning, announcing that many dikes had been broken and that soldiers on leave were ordered to report for duty.

On the roof of their farmhouse, Ria's parents were calm, soothing the children as the flood swept away their liveli-hood. On Sunday afternoon, the barn collapsed under the sheer weight of water pressing against its walls, drowning their horses, chickens and cattle, and filling the family with dread that the same would happen to the farmhouse. They spent Sunday night there and, miraculously, the farmhouse stood firm against the elements. That evening, the storm finally abated. The next morning, as the water level receded a little, Ria and her family clambered back into the house and sheltered on the first floor, waiting for rescue.

As in Britain, the flood hit towns and villages harder than it did big cities. While Rotterdam's harbour flooded on Sunday morning, the city remained dry behind its sea dike. The decisive reaction of mayor Jaap de Vogelaar meant Rotterdam did not face floodwaters from the breach at Nieuwerkerk aan den IJssel. The survival of the port was critical to the Netherlands' future. While much of Rotterdam's centre was still a flattened ruin after being smashed by Luftwaffe bombing during the war, commerce endured and it was still the world's third busiest port after New York and London.

Where the dikes were broken, roads and buildings remained submerged for days. When high tide returned, the flood rose again. The storm had brought death and destruction to Belgium too, where parts of Ostend and Antwerp were flooded, but the disaster was on a far larger scale in the Netherlands. Around 1,600 square kilometres of land was underwater, including nearly all of the province of Zeeland, and tens of thousands of people were stranded, many of them in villages and lonely farmhouses, like the Geluk family. Survivors clung to trees and rooftops without food or clean water. The tidal currents that swept back and forth through the breaches pounded against buildings and erased roads, causing further destruction and making it harder to mount a rescue.

Motorboats and rowing boats came through breaches in the dikes and across the flooded polders to ferry people out to larger ships waiting off the coast. Britain and the US provided helicopters to search for survivors and to winch people up from roofs – at the time, the Dutch military only had one helicopter of their own – while amphibious vehicles motored out to marooned villages.

Rescue came for Ria and her family on Monday afternoon, when a fisherman in a rowing boat reached the farm. For a few days, the group of survivors in the Geluk farmhouse had got by on half a pail of milk and a few buckets filled with drinking water. In the village nearest to the Geluk family, only two out of 20 houses survived intact. Dazed children arrived at shelters from the submerged villages of Zeeland, their faces filled with such blankness that worried mothers desperately tried to elicit a reaction from them by waving their hands in front of their faces.

Survivors searched for their families. The Geluks learned quickly that Ria's paternal grandparents had drowned when their house collapsed, but it took five more days for Ria's mother to learn from the Red Cross that her own parents had been rescued. Ria, her mother and sister were cared for by relatives living in the north of the country.

Her father stayed in Zeeland with the other men; there was work to do. Where the water receded, it revealed a shattered landscape of debris and splintered wood, smeared with mud. But the most urgent task was not rebuilding but gathering up, identifying and burying the dead. In Oude Tonge, Cor van de Tonnekreek, the boy who had wriggled across a pole to rescue dozens of fellow villagers, refused to be evacuated and stayed on to help guide soldiers who were searching for bodies. With Oude Tonge's cemetery under seawater, the only dry land available for burial was the dike, where all 300 villagers who died in the flood were laid to rest.

It took months to rebuild the country's flood defences. The last dike to be repaired was the one nearest to the Geluk family farmhouse. Here, the breach was so great – 200 metres across and 20 metres deep – that it was sealed with four reinforced concrete caissons, giant slabs of concrete that were 60 metres long, 20 metres wide, 20 metres high and hollow inside so they could float and be towed into position, before being filled with water and sunk to the seafloor.

These monoliths had originally been constructed to provide sheltered conditions for the D-Day landings on the coast of Normandy. Four caissons were towed across from Britain and submerged, to cheers from onlookers, at midnight on 6 November 1953.

It took another three months for the seawater to be pumped out of the polder behind the dike so that Ria's family could prepare the ground for planting. In the spring of 1954, they planted their first crop of wheat and barley. A year later, the family moved back into the renovated farmhouse. Ria's parents busied themselves with work, but there was never any discussion of what happened on the night of the flood. 'We had the Second World War and the floods,' Ria recalled. 'And people didn't talk about either one.'

A week after the disaster, Queen Juliana declared in a radio broadcast that the disaster was 'not caused by human corruption'. The queen, who toured the flooded islands and trudged through mud in a pair of rubber boots and headscarf, ascribed the disaster to fate, which could not be staved off by humanity. This was certainly a message the Dutch government wanted to convey; ministers had cut the budget for water management the year before the floods. But wind and tide alone were not to blame. Part of this disaster was man-made. The crisis had exposed a piecemeal approach to managing the threat from the sea, with responsibility divided between water boards and central government, and a failure to maintain dikes that had proven lethal.

No country will ever be completely immune from floods, but the question raised by 1953 was how far the Dutch were willing to go to ensure their safety. An engineer believed he knew the answer.

Johan van Veen was a patient man when he was studying tides or measuring currents and the drift of sand along a coastline, but he had little time for people he considered his intellectual inferiors – even when they were officially his managers. As chief engineer for a Rijkswaterstaat research unit from 1935 onwards, van Veen's intellectual abilities won him the freedom to study, measure and model the Dutch coast, but his reluctance to submit to hierarchy made life difficult, as his job required him not just to investigate but also to inform and persuade his managers of the need for action.

Van Veen's studies left him convinced that the dikes in south-western Holland were too low and too weak to resist a storm surge, yet his warnings to his superiors went unheeded. With just enough respect for his bosses to use a pseudonym, 'Dr Cassandra', he went behind their backs and published journal articles urging the government to fortify the coast.[7] But like the Trojan princess cursed with prophecy, there was little chance that anyone would act: the Great Depression of the 1930s meant there were long queues outside dole offices. People were pawning clothes or taking in sewing to make ends meet, and with Germany rearming and its leader making increasingly aggressive threats, the priority was guns, not flood defences.

In the spring of 1939, it seemed, finally, that van Veen was being listened to when the Dutch government appointed him to a new commission to investigate the risk of storm surges. This commission warned against the custom of building defences according to historic floods, saying that water could rise much further as sea levels were rising.

But a year later, the Germans invaded, overwhelming Dutch forces in days, and any maintenance work on the sea wall or further study of coast, tides and estuaries came to a halt. Van Veen was at a loose end and began work on a history of the Dutch struggle to win land from the sea, but as he did so, a new and more radical solution to his country's flooding problem was crystallising in his mind.

Raising and strengthening sea walls, as Dr Cassandra had urged, was the traditional response to the threat of flooding but van Veen contemplated an alternative. Instead of ever higher ramparts, what if engineering could reshape the entire country's coastline to slay the water-wolf forever?

CHAPTER 4

A MAN-MADE DISASTER

Cuchlaine King, a young academic at Nottingham University, arrived on the Lincolnshire coast soon after the 1953 floods as officials began searching for an explanation.[1] On this low-lying shoreline, 43 people had died and thousands had been evacuated. The sea had inundated the wooden chalets of Butlin's holiday camp at Ingoldmells, north of Skegness, where tens of thousands came to stay every summer, lured by the promise of 'a week's holiday for a week's wages', enjoying a ride on the Big Dipper, dances in the evening and tramping over the dunes for an invigorating dip in the North Sea.

King, a geographer and the daughter of a Cambridge geology professor, loved being outdoors, a breeze blowing through her long, curly hair as she hiked up glaciers or explored sand dunes in shorts and a pair of sturdy walking boots. The pleasure that drew her on was not the primal thrill of reconnecting to nature that attracted factory workers to Butlin's. Instead, she examined the movement of wind and sea across rock and soil, and the patterns made by people and animals, finding a story in the changing shape of the earth like a detective recreating the scene of a crime.

At Ingoldmells, King's exploration showed how people had forfeited the protection of nature. The dunes that protected the town are covered in marram grass, with distinctive spiky leaves that are perfect for children's games of hide-and-seek and dense roots that hold the sand in place. She discovered that the trampling of grass by holidaymakers' feet, as well as nibbling by rabbits, had allowed the wind to whip away sand and carve channels through the crests of the dunes. When the storm surge came, the sea poured through the channels, battering the dunes and sweeping into the low ground behind. The storm had struck with such force that, on some beaches, the sand and shingle had been ripped away, exposing the clay underneath.

The surge had devastated some places and spared others, and this was not entirely by chance. Skegness was protected by a wide beach, with high dunes which absorbed the force of the water. Between Mablethorpe and Sutton, two seaside towns on a short stretch of Lincolnshire coast where narrow dunes lay behind concrete sea walls, high water swept away dunes and walls alike, flooding the settlements. Sixteen people died in Mablethorpe and Sutton, while in Skegness, no one drowned.

Fittingly for someone who studied the Earth, King took a long view, knowing that, over the past six centuries, the Lincolnshire coastline had been forced back continually by the sea. Since the 14th century, five medieval parish churches had been engulfed by the waves between Mablethorpe and

Skegness, and an entire village had been wiped away. The -ness in Skegness was from the Old Norse word for a headland, a spit of land sticking out to sea, but the ness which gave the town its name had itself been swallowed by the water. Change was constant, but, with human intervention, the pace of transformation was shifting.

In her report on the disaster, King warned that the inflexibility of concrete sea walls often accelerated the degradation of beaches. Dunes provide a reserve of sand which is combed down by waves to reinforce the beach, but walls cut off this movement, leaving the beach thinner and more fragile. At high tide, some of the energy of the waves was transferred along the seaward side of the walls to undefended edges, attacking dunes along their flanks.

As King surveyed the beaches in the weeks after the surge, frantic activity to patch the gaps was underway. Eroded dunes were being shored up with towering walls of sandbags, while lorries and bulldozers drove great piles of chalk and rubble to fix breaches in sea defences. A remote and tranquil coastline had become an industrial scene, busy with the constant hum and grind of heavy machinery.

The inquiry that King reported to had been ordered by David Maxwell Fyfe, the home secretary. Nowadays, an inquiry is a near-automatic reflex for a government following a national tragedy, but 1953 represented one of the first occasions in British history that this had happened after a natural disaster. Britain in the 1950s was a far more religious society than it is now and some were inclined to see God's hand in the disaster. But 1953 also saw the first stirrings of a modern attitude, which saw disasters as man-made and asked who was to blame.

It was clear, from looking across the water at the even greater tragedy in the Netherlands, that Britain had suffered a near-miss. The combination of tide and surge, and the vulnerabilities in sea defences that came together to cause deaths on a large scale, could easily be repeated. Understanding what had gone wrong that winter was crucial to preventing an even greater disaster.

John Anderson, a brilliant scientist who chose a career in the civil service, was put in charge of the inquiry. Before the war, Anderson was given the task of preparing for air raids at a time when aerial bombardment of cities was regarded with apocalyptic dread. His name was given to Anderson shelters, semi-circular tubes of corrugated steel that were half-buried in the earth in families' back gardens to provide basic protection from the Luftwaffe.

After Germany invaded Poland, he was appointed home secretary and then, from the outbreak of the Blitz in 1940 until the Allied landings in Italy in 1943, he headed the Lord President's Committee, the body that coordinated Britain's wartime economy. He was lord chancellor for the final years of the conflict. After the war, he was raised to the House of Lords, as Lord Waverley, and the flood inquiry became known as the Waverley Committee.

The task of heading the flood inquiry suited an official who combined a background in science with wide-ranging administrative ability. But Anderson was also a man who posed little political risk. He was a long-serving insider with no affiliation to any political party and, for all his intellect, lacked an instinct for human behaviour.

Before the war, he had doggedly argued against building deep bunkers for civilians, insisting that effective shelters needed to be close at hand. When the bombing started, Londoners ignored him and headed for the Tube stations. Anderson was a technocrat who understood systems better than people and was therefore unlikely to deliver a stinging indictment of government failure.

Following a war which had been won with scientific breakthroughs, from radar to atomic weapons, the inquiry put expertise at its core, appointing a meteorologist, an oceanographer and a geographer to the committee, alongside senior civil servants, while calling in further evidence from engineers and scientists. Alongside King's report on the fragile dunes of Lincolnshire, a civil engineer, Sydney Mobbs, warned of the impact of human interference on beaches.[2]

Timber groynes, running down from the top of a beach into the zone where the surf breaks, became a popular feature of British beaches in Victorian times as a way of trapping the sand and shingle that would otherwise drift further down the coast. Holidaymakers huddled by them as shelter from rain and wind. On the bleak north Norfolk coast, large groynes had been built projecting out to sea to protect the beaches at both Sheringham and Cromer.

Mobbs had worked there, before and during the war, as engineer to the agency protecting the coast and he grasped that the groynes were causing immense damage to the coastline further south. Building large groynes to protect popular beaches had diverted some sand out to sea, rather than letting it be carried down the coast to feed other beaches. People had altered the balance of the coast and left part of it more vulnerable. South-east of Cromer, the principal line of defence against the sea was a 26-kilometre stretch of giant sand dunes and, just as they had done at Skegness, holidaymakers had trampled the grass holding the dunes together, leaving them weaker in the face of the onslaught.

In July 1953, the Waverley Committee delivered an interim report. This recommended the creation of a national weather warning system, addressing one of the main failings of the 1953 floods: the fact that, although the storm surge moved relatively slowly down the coast, communities further south were taken by surprise. Waverley delivered his final report in the summer of 1954, a moment of optimism as the last of wartime rationing ended and British youth acquired a taste for the frantic energy of rock'n'roll.

The report did not assign any blame, and there is no mention of the memo that, the summer before the floods, imposed a halt on all national defence works. The committee identified the natural causes of the flood – the storm surge and the tide in combination – and noted that these had different sources. The storm was atmospheric, driven by the fierce wind from the north-west, but the tide was astronomical, driven by the sun and the moon.

The committee's investigation found that the flood did not occur at the peak of the spring tide and that the surge, by itself, was not the highest on record; it was the combination of the tide and surge that produced a record-breaking wall of water. The storm whipped the sea to more than 5 metres above mean sea level along an 80-kilometre stretch of the Lincolnshire coast, from the Humber estuary down to the beaches south of Skegness. The sea rose higher than this at points along the Norfolk coast, passing 6 metres at Blakeney.

Critically, Waverley's experts found that it could have been worse. If the wind had generated a surge to match the highest on record, and this had coincided with the peak of the spring tide, the water level would have been 2 metres higher still. There was one more stroke of fortune; the rainfall inland was below average for winter, so the surge of seawater that funnelled up the Thames did not encounter a swollen river coming down. While disastrous for the people living there, the inundation of Canvey Island, along with marshland in Essex and Kent, had helped reduce the surge that hit the capital.

The cost of protecting the entire east coast against the worst conceivable flood could not be justified, the committee decided, but they drew attention to the need to defend London. A large part of the capital is built on what was once marshland and is below high tide, while the Thames has been straightened and dredged over the centuries, making it faster and deeper. The vulnerability of the Tube to water had been demonstrated during the war, when the Luftwaffe dropped a giant bomb on Balham underground station on a winter's evening in 1940. The bomb itself did not directly inflict casualties, but the explosion punched holes in water and sewage mains, flooding the station where hundreds sheltered and drowning 68 people.

The surge of 1953 had lapped the top of the parapet along the Victoria and Chelsea embankments in central London. One way to meet an even higher flood was by raising the city's defences, but this would require the rebuilding of the districts alongside the river, moving and relocating the roads, railways and warehouses by the Thames. Another way to protect the city was to build a barrier. The idea of a barrage across London's great river had been debated for centuries – alongside other engineering fantasies, such as a tunnel under the English Channel – and it had gained impetus after a previous storm surge swept up the Thames in 1928.

Britain still had a global empire in 1953, governing 38 colonies and protectorates from Kenya to Hong Kong, and while the US had surpassed it as an industrial power, British engineers had a legacy of transformational projects both on their home islands and around the planet. They had carved railways through West Africa's lush tropical forest, created giant irrigation projects in Sudan and the Indus Valley in south Asia, and, by 1954, were completing the construction of a dam on the Nile in Uganda.

Bridges and rail lines sped the movement of imperial troops and the extraction of palm oil, rubber and tin. Irrigated fields grew cotton and indigo for export. Britain's engineers were, literally, empire builders. London was also the world's second busiest port after New York and the need to keep the river free from obstruction clashed directly with the idea of a defensive barrier on the Thames. Before the war, the Port of London Authority had scuppered the idea of a barrage by raising the prospect that it could be bombed by enemy aircraft to release a deluge – the drowning of Balham Underground on a giant scale.[3]

Waverley's report acknowledged that the climate was changing, with the implication that the worst disasters of the past were no guide to the future. The inquiry noted the steady increases over the past 100 years in the high-water levels reached by exceptional storms, referred to the melting of polar ice and the shrinking of glaciers, and combined these observations with the fact that London and southern England were sinking as Britain adjusted to the retreat of a massive ice sheet at the end of the last Ice Age.

The report predicted a rise in sea level relative to land of 'a foot, or even more, a century'; that is, 30 centimetres in 100 years. The melting of ice that Waverley referred to is likely to have been caused by a natural weather pattern and was nothing like the change that humanity is now inducing in the planet, which will see an accelerating rise in sea levels. Our current understanding is that the seas around the UK have risen by a little over 15 centimetres since 1900 and that sea levels off the British coast will rise by a further 50 centimetres – and perhaps as much as 80 centimetres by the end of this century.

With just half a metre of sea level rise, around 200 kilometres of England's present-day coastal defences would be vulnerable to failure. Losing more of the kind of natural protection that Cuchlaine King identified – sand dunes, shingle beaches, salt marshes – will increase the risk.

*   *   *

The link between burning fossil fuels and climate change was not common knowledge in the 1950s, but scientists had already made the connection and, indeed, some regarded it as a welcome prospect. Greenhouse gases, such as carbon dioxide and methane, are naturally present in the atmosphere, fuming up from venting volcanoes, wildfires and decomposing vegetation, and they have made Earth habitable. Without those gases trapping some of the sun's heat in the atmosphere, like the glass of a greenhouse shielding the entire Earth, the planet's temperatures would be sub-zero. This 'greenhouse effect' of carbon dioxide was first demonstrated by Eunice Foote, an American scientist, in the mid-19th century.

Early on, some thought that further planetary improvement would come from humanity releasing the energy trapped in coal and other ancient plant matter. Svante Arrhenius, a Nobel Prize-winning Swedish physicist, calculated in the late 19th century that industrial consumption of fossil fuels would result in a temperature increase of between 3 and 4° C, so that 'our descendants … might live under a milder sky and in less barren surroundings than is our lot at present'. Writing in 1938, the British engineer Guy Callendar acknowledged that few were prepared to admit that man's activities could have any influence on a vast phenomenon like the making of the climate.[4]

But, Callendar said, this was not only possible, 'but is actually occurring at the present time'. He documented the upward trend in Earth's temperatures in the early decades of the 20th century and linked this to the combustion of fossil fuels. Again, he thought this beneficial, at least for farmers in the northern hemisphere, adding that 'in any case the return of the deadly glaciers should be delayed indefinitely'. Callendar's work was dismissed by other scientists. The director of the Met Office gave him the back-handed compliment of praising the 'amount of work' the engineer had put in, before rejecting the link between carbon dioxide and increasing heat as a coincidence.

By May 1953, just a few months after the North Sea floods, the Canadian physicist Gilbert Plass revived the theory that man was altering the climate in a speech to a conference of scientists in Baltimore. Plass expanded this initial report into a scientific paper published in 1956, which sounded the alarm.[5] Rather than milder skies in northern Europe or bumper harvests of wheat in the Arctic Circle, Plass warned that the quantity of carbon dioxide being pumped out by industry was sufficient to upset the balance of the climate. It was far more than could be removed by the growth of plants and absorption into the oceans. His calculations predicted a temperature change of 3.6° C if the amount of carbon dioxide in the atmosphere doubled. 'A relatively small change in the temperature can have a large effect on the climate,' Plass wrote.

The missing piece of the puzzle was to establish just how much carbon was spewing into the air. This was supplied a few years later by Charles Keeling, a young American researcher who loved the outdoors. Keeling began taking measurements of carbon dioxide in California's Big Sur region, a belt of redwood, conifer and oak forest where the Big Sur river empties into the Pacific. In 1958, Keeling took his measuring instruments to an observatory more than 3 kilometres above sea level on the flank of the world's biggest active volcano, Mauna Loa in Hawaii, a location chosen because of its remoteness from cities and industrial activity that could interfere with the results.

His first reading here came in at 313 parts of carbon dioxide per million. Taking regular readings from Mauna Loa, as well as from research stations in Antarctica and California, Keeling established there was a steady increase in the concentration of CO2 in the air. By the time Keeling died, in 2005, it had reached 380 parts per million. The quantity of CO2 is still a tiny fraction of the mix of gases in the atmosphere, but it's enough to have significant effects.

A civilisation powered by fossil fuels is now altering the planet's climate and creating a world that's hotter and more deadly for humans. A report by the Intergovernmental Panel on Climate Change, the UN body that advises on the impacts of a warming world, says that some of the changes humanity has set in motion – such as sea level rise – will be irreversible for centuries. The IPCC warns that extreme events such as the 1953 floods, that previously occurred once in 100 years, could happen 'every year by the end of this century'.

It was not just the action of wind and waves that made 1953 so disastrous. The destruction of life and property was 'unprecedented', according to Cuchlaine King's research on the Lincolnshire coast, because so many people had settled in the most dangerous zone between the sand dunes and a sea defence known as the Roman Bank, a clay embankment which runs along the coast from Skegness to Ingoldmells. Making the link between the processes of the natural world and the role of human nature, she found that the deterioration of the beaches had been compounded by flawed choices. 'It is unfortunate that the holiday industry has attracted a large population into the most dangerous area near the coast,' she wrote.

King argued that the best defence against flooding was one provided by nature, 'a broad belt of high, well-vegetated dunes, fronted by a wide, high beach'. But she acknowledged that this would not always be possible and that sea walls would be essential for the safety of the coastal population, even if they were detrimental to the growth of dunes. As Britain turned to rebuilding after the floods, King's words were largely ignored. Instead, artificial defences were repaired and extended. Since the end of the Second World War, the British coast has lost around a fifth of its natural protection as environments like salt marshes were eroded by flood walls.

Concrete walls reinforced with steel were built to replace the breached defences around Canvey Island, further raised and strengthened over the years to create a 14-kilometre ring of armour, more than half a metre thick in places. Two of the creeks that extended the sea's fingers into the town have been dammed. While building societies hesitated, the council offered mortgages directly. In the three decades after the flood, Canvey's population trebled, from around 11,000 to 36,000. Six years after the floods, a storage terminal on the island received the world's first seaborne cargo of liquefied natural gas, as the oil and gas industry that was heating the planet helped revive the town's fortunes.

Sea walls were restored along the coast and new walls built where sand dunes had been obliterated by the storm. This was, in part, driven by necessity. Sand dunes are repaired by the wind blowing more sand onto them, but this process takes time and a single summer of natural repair would not be enough to restore the damage done.

Alongside these walls, a network of tide gauges was set up, from Stornoway in the Outer Hebrides to Southend on the Essex coast, giving readings that can provide warnings up to 12 hours before high water arrives. Emergency sirens issuing the same uncanny wolf-howl as an air-raid alarm were installed to provide warning of imminent flooding.

Creekmouth, the tiny London neighbourhood that was flooded, would never have anyone living on it again. The homes there were demolished, the people rehoused and it became an industrial park. But Creekmouth had never been a large settlement; no more than a few dozen cottages built for the employees of a chemicals company, it was the exception. All along the eastern edge of England, officials in charge of coastal protection built fortifications against the tide and the population swelled, growing much faster than it did inland.

By the spring of that year, the long shadow of the war was lifting; a few days after the floods, the government had ended chocolate and confectionery rationing, sending hordes of giddy schoolchildren into sweet shops. At the end of May, a British-led expedition conquered Mount Everest, with the New Zealanders Edmund Hillary and Sherpa Tenzing Norgay standing at the highest spot in the world for 15 minutes. Four days after this reassertion of human dominion over nature, a young queen was crowned in Westminster Abbey.

The genesis of the Thames Barrier is a revealing parable about humanity's ability to act in the face of disaster. The Waverley Committee's recommendation of a technological solution that would close the river at times of high water was unprecedented in one of the world's great port cities. In the past, British governments had always relied on raising the height of earth embankments and river walls of brick and concrete for London's flood defence. As global shipping boomed after the war, the capital's docks experienced the busiest years in their history. In the days before shipping containers, that meant strenuous physical labour for gangs of dockers who packed the holds of ships. Port officials feared that waiting to navigate a barrier would impose severe delays on shipping. It took a charismatic scientist, and a powerful fright for Britain's ruling class, to propel flood protection up the political agenda.

On a Friday in early December 1965 – more than 12 years after the terrible floods on the east coast – an exceptionally high tide surged up the River Thames and came close to flooding the Houses of Parliament. As water seeped through the parapet wall on the river terrace of the Palace of Westminster, Tom Driberg, a former gossip columnist turned Labour MP, interrupted a Commons debate to warn fellow MPs that 'at this moment, down below, the River Thames has reached a level unprecedented in living memory'. The water was within a few inches of pouring into the chamber, Driberg added, and it was still nearly an hour until high tide.

The deputy speaker, Roderic Bowen, an army captain in the war, calmly assured MPs that the situation was being watched and the debate went on. But parliament had been taken by surprise and it was sheer luck that Westminster escaped being inundated; the water stopped just 20 centimetres short of the wall that protected the terrace. The high tide that day reached nearly 7 metres, more than a metre above the level predicted by forecasters.

The proximity of the threat concentrated minds and a fresh report on the Thames was commissioned from Hermann Bondi, a brilliant mathematician and cosmologist whose large spectacles, receding curly hair and faintly Austrian-accented English gave him the appearance of an eccentric scientist from a children's book. Bondi was an Austrian-born Jew who left 1930s Vienna as a young man and studied mathematics at Cambridge.

The experience of watching his former homeland collapse into virulent anti-semitism marked him. At school in Vienna, another boy remarked, casually: 'I haven't got anything against you personally, but when we come to power we will have to get rid of people like you.'[6] In March 1938, the year after Bondi arrived in Britain, Germany annexed Austria. He cabled his parents, advising them to leave the country immediately. Both of them did, heading for Switzerland and eventually settling in New York. Bondi's mother left Austria first, but his father escaped just hours before German troops arrived. For his family, at least, disaster had been averted through decisive action.

Throughout his life, Bondi was a man who cut through waffle. A secretary taking notes during a committee meeting observed that Bondi wrote nothing down on the notepaper in front of him except a string of numbers.[7] 'When everybody had had their say, he would say: "Ladies and gentlemen, let me summarise." And then he would expertly sum up what happened in the discussion,' the secretary said. 'Everybody felt they had their say, but he always got the outcome he was perhaps looking for.'

Bondi found that the height of flood tides in London had been rising over the past 150 years. His report in 1966 warned that if the London Underground was penetrated by floodwater, it would take months to put the system back into action, with tunnels filled with water, mud and debris, and electrical equipment rendered unusable. It would deal, he said, a 'knock-out blow to the nerve centre of the country'. The disruption caused by a surge that exceeded the 1953 level might exceed the effects of the Blitz, Bondi told ministers. Breaking the stalemate between the safety of the capital and its commercial interest in staying open to shipping, Bondi recommended the construction of a tidal surge barrier on the River Thames.

Waverley's report had suggested placing this at Long Reach, near Dartford, where the river runs straight, making it easier for ships to navigate through a barrier's openings. But Bondi advised placing a barrier further upstream and closer to the centre of London, at Woolwich or just upstream of

West India Docks – the area now known as Canary Wharf – as it would reduce costs to use a spot where the river is narrower. The choice of location does not just matter for shipping. A barrier placed closer to the sea extends protection to towns and villages in Essex and Kent, downstream of London, while placing the barrier closer to the capital means communities living further downstream must look to their own defences by shoring up flood walls and embankments.

Mary Kendrick, a geographer at the government's Hydraulic Research Station in Oxfordshire, led a team that experimented to discover the impact of a barrier further downstream.[8] Like Cuchlaine King, Kendrick won the respect of her largely male counterparts for being unfazed by a physical challenge. She undertook fieldwork in survey vessels on the Thames in thick jumpers and corduroy trousers, then conducted further research on a 115-metre-long and 12-metre-wide model of the river in a shed in Oxfordshire. The team sent miniature tides down the concrete riverbed of their model that reproduced the bends of the Thames from the west of the city to the sea. A second model mapping a shorter stretch of the river included tiny replicas of the turrets of Tower Bridge, a decorative touch that was redundant for scientific purposes but offered a pleasing visual shorthand of the city under threat.

There were two significant risks. The first related to the tidal nature of the Thames. Twice a day, the salt water of the North Sea sweeps inland on the flood tide and then retreats on the ebb tide, the river level rising and falling by up to 7 metres. As the sea floods in, taking two and a half hours to sweep up from Southend in Essex to Teddington on the western edge of London, it brings silt with it, tiny particles of matter finer than grains of sand that have been chipped away from the coast and borne along on the water. The fresh water of the ebb tide washes much, but not quite all, of the silt out again. A barrier in the river risked slowing the flow of water and creating a build-up of silt.

This might not sound like much of a hazard, but the construction of a barrage across the River Eider in Germany in the 1930s had succeeded in protecting its higher reaches from flooding while choking its lower reaches so completely with silt that the rest of the river disappeared.

The proposed Thames barrier posed a further threat to those living downstream of it – the risk of a 'reflected wave' sweeping back down the river when the gates of the barrier closed to protect the capital. Kendrick and her colleagues examined three potential locations: at Woolwich on the eastern flank of London, at Silvertown a little closer into the city, and at Blackwall, even further into London.

Their experiments found that if the barrier was only closed to exclude surges and prevent flooding, the structure could straddle any of these points without disturbing the equilibrium of the river. But if the barrier was used on a regular basis to manage the tide and exert a tighter human grip over the river, it would transform the Thames, reducing its depth by silting it up. The further east – and thus closer to the sea – the barrier was placed, the bigger the effect would be.

The researchers' experiments on their miniaturised river models showed the impact of closing the gates on towns further downstream. Here, the crucial factor was timing. If the gates were closed at low water, the effect was negligible, raising downstream levels by fewer than 30 centimetres at most. But if the gates were closed after the flood tide had gained momentum, water levels downstream could rise by more than 70 centimetres. Officials in charge of the barrier would need to identify an incoming surge early enough to begin closing the gates long before high water arrived.

As Kendrick began her research in the late 1960s, London planners had devised another flood response: the first housing development in a modern city that took flooding into account. In July 1968, Terry Gooch and his family were collected from Peckham in a chauffeur-driven limousine to become the first residents of a futuristic new estate, Thamesmead. The area, once known as Erith marshes, had been grassland on the south bank of the Thames where horses grazed and it had been submerged in the surge of 1953.

Council architects designed a flood-proof utopia, blending 13-storey residential towers with low-rise maisonette blocks raised above ground level, with garages underneath the living quarters and elevated walkways providing escape routes. The new name, Thamesmead, was chosen in a competition and hinted at its rural past and waterfront location, but the style was pure modernity: bare concrete surfaces and straight lines. Utopia quickly turned sour. Thamesmead never had a rail link to central London, nor a bridge across the river, and felt isolated, while the walkways became dark corridors at night, haunted by the fear of crime. Its brutal, swaggering beauty made it the perfect setting for Stanley Kubrick's film of violent youth, *A Clockwork Orange*. But it was one of the first attempts in modern times to live with rising water, rather than emphasise holding it back.

Rivers and seas sometimes meet the land in a sharp line – the edge of a cliff dropping away to deep water. But there are many places where the transition is more gentle: a wetland where earth and sea mingle or a gradually sloping beach leading to a shallow bay. Humans have always been an amphibious species, delighting in paddling canoes through creeks, hunting ducks in marshes or dipping into the sea for a swim. Embankments and flood walls provide protection from the threat of inundation, but they have also imposed straight lines where none existed before, turning the broad and gentle Thames of the 17th century into a faster and deeper river, with cliff-like embankments that separate people from the water.

By 1974, when construction of the Thames Barrier began, merchant shipping on the upper stretches of the river had gone into decline as cargo sped across the oceans in containers, with ships needing deeper water and fewer men to load them. Work became scarce in the East End docklands and the British Empire was unravelling; vast colonies in Asia and Africa were now independent and all but a handful of scattered islands – the Falklands, Hong Kong – followed them. Although no longer a world power, Britain still had the resources and technical ingenuity to build a shield against the storm.

Two firms of consulting engineers were appointed and they differed over the design, arguing for the rival merits of gates that lowered into the river or a barrier with arms that swung out across it. The final design of the gates was an inspiration that struck Charles Draper, a designer working for one of the engineering firms, when he was twiddling gas taps on a fire. It made an ingenious use of the space available. Nine steel piers would be installed in the chalk bed of the Thames at Woolwich, around 13 kilometres downstream of St Paul's Cathedral, spanning a 520-metre width of the river. Across the main part of the barrier, concrete shells were dug into the riverbed, with the idea that the six steel gates would lie flat in these when they were not in use.[9]

When the barrier needed to close, these steel gates would be pulled up by hydraulic arms, rotating them from lying flat and flush with the river bed until they were vertical and their curved exteriors faced the sea's incoming tide. Two of these rising gates would span 31.5 metres, while the four central ones would each control a space that was 61 metres wide.

The gates were being constructed as the Irish Republican Army waged an intensifying campaign of attacks on the British mainland, beginning in 1973 with car bombs at the Old Bailey and in Whitehall, and with the size of devices used and the devastation caused escalating as the decade went on. Late in the evening of 17 January 1979, a bomb punched a hole, 18 centimetres in diameter, in an aviation fuel tank at an oil terminal on Canvey Island. The jet fuel spilled, but did not ignite. At midnight, a caller gave a warning and police were alerted to a second device at the Blackwall Tunnel beneath the Thames. As officers searched the tunnel, this second bomb went off at a gas storage container near the tunnel's south exit, sparking a fire and igniting a second gasholder.

While no one was killed or injured in these explosions, the attacks underlined the vulnerability of Britain's critical infrastructure. The Thames Barrier was designed to be resilient, with gates of thick steel and piers of reinforced concrete. The barrier has three independent sources of electrical power, one from the south and two from the north. All three have failed simultaneously, just once, when an exceptionally powerful storm hit southern England in 1987. If grid power is cut off, diesel generators provide a back-up, with sufficient fuel to operate the barrier for seven days.

Margaret Thatcher visited the barrier in February 1979 when she was still leader of the Opposition in the House of Commons.[10] In a hard hat, she bustled approvingly around the site, climbing up a ladder to explore the building work and admiring the courage of divers who plunged into the cold and murky water to lay explosive charges on the riverbed or to move steel joists into place. She asked questions of workers in the firm, more powerful voice that a theatrical coach had helped her develop. One of the men, towering over her, gave a little bow and said: 'It's a pleasure to meet the next prime minister.' She was elated.

'I haven't seen any signs of industrial action,' Thatcher told a reporter confidently. 'In fact, I have been very impressed by what I have seen.' Britain was still in the grip of an unusually long and bitterly cold winter. This, combined with industrial action, made the country feel increasingly chaotic; there were strikes by car workers, railway workers, truck drivers, hospital cleaners and even gravediggers. The Tories had been running a poster campaign with the slogan 'Labour isn't working', alongside an image of a long queue waiting to enter an unemployment office. The humiliating words and image had the sting of truth.

Government planners became preoccupied by the prospect that a storm surge would hit the capital before the barrier was in place. Flood contingency plans drawn up after Thatcher became prime minister in 1979 involved relocating the staff working out of 10 Downing Street and moving the PM to her country residence, Chequers. But Thatcher, as it turned out, had dismissed the threat of strikes at the barrier too soon.

The giant gates, the biggest of which were 20 metres high and weighed more than 3,000 tonnes, were manufactured in Darlington and assembled at Port Clarence, a village on the River Tees in north-east England, before they were loaded onto barges for the sea journey south. Early in 1982, industrial action by dock workers on Teesside held up delivery of the gates. Peter Walker, the agriculture minister, told Mrs Thatcher he feared a critical delay in completing the defences and a potentially dangerous increase in the flood risk that coming winter. Walker raised the question of whether the army should be brought in to load the gates.

Official records, kept secret for 30 years, reveal that Thatcher considered a hazardous emergency plan for London if the gates were not delivered on time. Sea walls were being raised further down the Thames, but this could make matters worse in London, funnelling a higher surge tide upstream. Civil servants advised the prime minister that if the barrier was not ready by winter, it might be necessary to breach flood defences further downstream in Essex and Kent and sacrifice low-lying land there to protect London.[11]

Explosive charges would have to be set in advance so that they could be detonated to punch holes in the downstream defences when a surge tide that threatened the capital was on its way. This would, of course, be disastrous for towns closer to the sea. The officials warned Thatcher that this desperate plan meant a 'major risk of loss of life' on Canvey Island, where there would be insufficient time to evacuate ahead of the surge. In blue ink, the prime minister under-lined these words.

The memo paints an apocalyptic picture of devastation in the capital. The officials forecast that, without the barrier in place, there could be 'hundreds rather than dozens' of casual-ties in London, with buildings collapsing and people being swept down open manholes on flooded roads. The water was likely to pool and could take a week or more to drain away from some neighbourhoods. Floodwater would be polluted by sewage and there would be widespread electricity failure and damage to roads. The surge would cause earth slippages that would fracture sewers, gas and water mains.

Using the army to break the strike was politically fraught, the officials warned, saying that if that happened, the barrier's workforce was likely to down tools in sympathy with the dockers. In the end, the choice between London and the counties downstream never had to be made as the strike was resolved and the gates delivered, but it highlighted the polit-ically sensitive calculations that were going into flood protection.

The barrier became operational in December 1982 and was first used two months later, on the night of 1 February 1983, when forecasters warned of a surge coinciding with high tide in the early hours. The first closure of the gates 'worked like a charm', one of the city council's engineers told a reporter. For the first time in history, London and the upper reaches of the Thames were cut off from the sea by a wall of steel. When Queen Elizabeth II officially opened the barrier in May 1984, she described its eight years of construction as a 'race against the tide' and acknowledged those who had died in the 1953 floods when, she said, the capital had narrowly avoided catastrophe.

In the long contest between humanity and nature along England's eastern seaboard, man now appeared to have the upper hand. Fittingly for a nation that had just emerged victorious from a titanic war, the response to the 1953 floods was militaristic, emphasising defensive lines and technologic-al solutions to the threat posed by water.

By the time the Thames Barrier opened, the map of the world was speckled rather than painted pink; the wharf that once unloaded rum and sugar from Caribbean plantations had become a financial district. The barrier represented a final triumph of imperial engineering.

The shock of 1953 and the high water threatening the Houses of Parliament in 1965 had impelled Britain to act. The growing awareness of the threat to the capital turned coastal protection from a neglected duty to a concern that preoccupied ministers.

On the other shore of the North Sea, the catastrophe of 1953 would set in motion an even more ambitious attempt to engineer nature. In the Netherlands, the case for action was even more compelling. The Dutch nation is a scrap of land between salt water and rivers, with more than a quarter of its terrain below sea level. Yet here, as in Britain, it required the intervention of a single determined individual to set the country on a new course.

CHAPTER 5

SEALING OFF THE SEA

Johan van Veen heard the rare warning 'dangerously high water' in the radio weather forecast and understood that the storm was a dagger pointed at the heart of the country. He did not sleep another wink that Saturday night.

The quick-thinking and courage of mayor Jaap Vogelaar and skipper Arie Evegroen had plugged one breach on the Hollandse IJssel river. But the first radio reports van Veen heard on Sunday morning confirmed that the surge sweeping up the Hollandse IJssel had broken through the dike by another village in central Holland, Ouderkerk aan den IJssel.

The engineer knew that the dikes in the centre of the country rested on a sinking layer of soft peat and had needed heightening many times since they were first raised seven centuries before. In the south west, Zeeland was already lost, but the flood threatened the cities of Amsterdam, Rotterdam and The Hague, where five million lived. When his daughter Anja walked into the room early that Sunday, he turned bleary-eyed from the radio and told her: 'Now, you'll see, they will approve my plans.'[1] But, first, he wanted to direct the rescue for himself.

Van Veen rushed to Ouderkerk after persuading one of his managers to dispatch a government car. The driver struggled to maintain control as sheets of rain seemed to sweep in horizontally. The breach he found there was 40 metres wide and, he thought, the river cascading through it looked like Niagara Falls. The field behind was already flooded, a little church at its edge partly collapsed. The men of the village were frantically piling sandbags into the breach, but the torrent seemed impossible to subdue.

At noon, a party of soldiers arrived and the water level in the river dropped as the tide turned. Joining forces with the farmers, the men raced to seal up the gap with van Veen guiding them, before the water rose again. At 4 p.m., the breach had been closed. 'It is truly incredible that central Holland still exists,' the engineer wrote to a British friend.[2]

Then aged 59, van Veen had suffered a serious heart attack five years earlier. This had done little to dim his appetite for work. He was a genius, capable of intuitive leaps from one branch of science to another. In an era before computers, the movement of tides was worked out with laborious human calculation. Realising that the flow of water back and forth in a tidal estuary was like the flow of electrons in an alternating electric current, van Veen had outlined how tidal channels on the Dutch coast could be simulated by running current through copper wire.[3]

His insight had the potential to save hours of time. But his mathematical colleagues, jealous guardians of their field, were dismissive of the breakthrough.

His heart attack had done nothing to quench the fire in him either. He was given to storming out in a fury, slamming the door on his way, when he felt other people didn't understand him – whether that was his wife Henny or his colleagues at the Rijkswaterstaat. His marriage suffered, too, from his addiction to work and from his infidelity.

Van Veen was a practical man whose brain and hand worked in unison and who could seemingly turn his dexterity to any craft. He once made his daughter a tortoiseshell bracelet and carved her name into it in elegant script. Another time, he invented a device for sampling sediment, the 'Van Veen grab', a manually operated stainless-steel claw that springs shut like a mousetrap around a handful of river sand or pebbles before hoisting it up to the surface. A machine of elegant simplicity, it is still in widespread use today.

He was never happier than when out on a riverbank, walking barefoot over mud and through stands of beachgrass with his trouser bottoms rolled up above his knees. Occasionally, he would take his children out of school for the day to come with him and see for themselves the Dutch art of water mastery. Experiments, whether with copper wires or models of water channels, were useful, but nothing substituted for direct knowledge of the river. River research, he wrote, 'comes first and last'.[4]

In 1933, a time when stormtroopers marched through Berlin to hail Germany's new leader while governments everywhere struggled to find an answer for the millions who were out of work, van Veen won the favour of a senior colleague. Recognising the engineer's drive and intelligence, the head of the Rijkswaterstaat put him in charge of a new study department, with a licence to explore at will. Granted a boat and a small team, van Veen was in his watery element, spending happy days on the coast taking samples of sand and measuring the flow of currents.

His measurements quickly led him to the conclusion that the dikes needed to be raised and strengthened – especially in the south west of the country – but a more radical idea was forming in his mind: that of closing the coast altogether.

This idea was inspired by history. The devastating St Lucia's flood of 1287 had altered the geography of the Netherlands, creating the Zuiderzee, that inlet of the North Sea that reached into the heart of the country. It was a haven for migratory fish, with shoals of herring and anchovy coming to spawn in its sheltered waters each spring. Since the 17th century, the Dutch had daydreamed of turning its brackish water back into a lake and taking back the seabed to make wheat grow where herring laid their sticky eggs.

The Zuiderzee had given Amsterdam's merchant princes a pathway to the sea – and the oceans beyond it – for their trading expeditions to Asia's spice islands. The sandbanks and shallows that created nurseries for fish made it treacherous for sailors; trading vessels returned with full bellies, sitting deep in the water and hundreds of vessels foundered when sudden storms blew up, including a British merchant ship from the time of Queen Anne that was recovered three centuries later with bottles of wine still intact. By the 19th century, the shallow sea was no longer needed for trade as a canal provided a shorter and safer passage from Amsterdam to the North Sea.

The inlet often flooded villages on its banks and a particularly devastating flood in 1916, which submerged the countryside north of Amsterdam and claimed more than 50 lives, rallied public support for a plan to close it off. The Afsluitdijk – the 'Enclosing Dam' – took five years to build. At its base, the builders placed mattresses woven from willow branches, ballasted with boulders, a traditional technique to slow the current's velocity and prevent it from eroding the dam.

A mix of rock and clay dredged from the bottom of the inlet and dumped into the water from barges made up the body of the dam. Every time the tide flooded through from the sea and ebbed back again, it generated a fierce current in the narrowing gaps, washing away the rock and clay as the workmen piled it up. As the breaches in the dam narrowed, the wash of the current between the gaps grew increasingly turbulent, so the barge captains struggled to position their boats in the right place. The work became focused on periods of 'slack water', the periods when the tide is turning and the current is least powerful. The threat of a violent storm, which might sweep away months of labour in a single night, hung over the engineers throughout.

In 1932, men and ships managed to outrun the tides, dumping a last load of rock and clay into the final breach and a 32-kilometre-long barrier sealed the Zuiderzee, like a cork plugging a gigantic bottle. The inlet had been split in two, with a saltwater inlet in front of the dam and a vast artificial lake behind. Grass was planted to bind the clay of the dam together and a road was laid across the top; a Heineken beer truck was the first freight vehicle to trundle across. Behind the enclosing dam, work began on secondary dikes, pumping stations and sluices to turn the shallow sea into dry land.

It is little known outside the Netherlands, but the size of this reclamation makes it one of the greatest feats of water engineering in world history. Across decades, nearly half of the area that was once underwater became land that's home to more than 400,000 people today. The rest turned into a freshwater lake, the IJsselmeer. A 2-metre-high bronze relief placed in a lookout tower at one end of the dam depicted three stonemasons busy laying blocks, with the boast: 'A nation that lives, builds its future.'

Thrilled by its size and ambition, van Veen dared to dream on the grandest of scales. If it was possible to build a 32-kilometre dike in the tidal waters of the Zuiderzee, it might be feasible to fortify an entire country. During the Second World War, the engineer worked on a plan to link the islands of southern Holland with barrier dikes and, in 1942, proposed sealing off the entire Dutch coastline. His idea, the *Verlandingsplan*, would involve manipulating the flow of the great rivers, the Scheldt, the Meuse and the Rhine, that flow down from France and Germany to reach the sea in the Netherlands.

The rivers – in other places, natural dividing lines that become the boundaries of empires and the places where armies clash – become harder to distinguish by the time they

approach the Low Country. From the viewpoint of water
birds circling above, the waterways of the western
Netherlands are flat and broad, interspersed with islands,
thickets of trees and swathes of marshy ground.

Long before the rivers finally empty out, the sea arrives to
give them a salty kiss; twice a day, the tide flows in and out of
the river delta, washing in the plankton that feeds oysters and
mussels. The rivers branch and intermingle on their way out
to join the salt water, and – as if to acknowledge this natural
confusion – the Dutch give the tangled channels that peel off
from the Rhine many different names: Nederrijn, Kromme
Rijn, Waal and Lek.

Van Veen's idea was to shift their flow so that their mouths
naturally silted up and they no longer emptied out into the
sea. His Verlandingsplan would take two centuries to
complete, but when it was done, the inlets where tides swept
in and out – and storm surges could penetrate deep inland –
would be blocked off permanently.

Van Veen's colleagues largely ignored his warnings about
the state of the dikes and found his notions about the coast
fantastical. In the autumn of 1942, as the German occupiers
tightened their grip, a new chief engineer was appointed at
the Rijkswaterstaat, Willem Harmsen, who sidelined van
Veen. Harmsen bluntly told van Veen that he lacked
self-awareness and leadership qualities, and took away his
supporting staff.[5]

The German occupation had made the work of maintaining sea defences increasingly challenging. The entire coast
was barred to the Dutch authorities, while German officials

demanded lists of Rijkswaterstaat employees who could be deported to work in the Reich. Some water defences were still maintained and strengthened, as the Germans did not wish to see their new province destroyed by flood. Work to reclaim land in the former Zuiderzee continued.

But, by 1942, nearly all the projects that the German occupiers didn't regard as valuable to their war effort were terminated. The Germans were pushing for the surrender of the ministry's vast stockpile of iron to fulfil Hitler's desire for thousands of kilometres of fortifications along Europe's western edge – an Atlantic Wall to hold back the Allies.[6] Heads of department at the ministry refused to fill out lists of names and the Rijkswaterstaat leadership resisted demands to hand over their iron. But such resistance was ultimately futile.

At a dark time, both personally and for his country, van Veen reassured himself that his plans were fundamentally sound. With the reserves of confidence that had won him so much loathing, he wrote in his diary that he knew more about water management than anyone else and that 'knowledge is power'.[7] But, at the time, it must have seemed that power was something else, something simpler: men with guns. Dozens of Rijkswaterstaat employees were forced to move to Germany and the iron stockpile was seized.

Unable to carry out research off the coast in the final years of the war, van Veen poured his energy into writing a book. Instead of an engineer's treatise, he wrote a colourful history of Dutch defence against the sea, stretching back centuries to the first dikes in the Middle Ages. *Dredge, Drain, Reclaim* was published in 1948 and celebrated the efforts with which engineering had subdued geography, each step more ambitious than the last.

As the Allies advanced through western Europe in the final stages of the war, the retreating Germans left ruins behind, blowing up bridges and breaching dikes to flood polders, as well as laying hundreds of naval mines and scuttling ships to block waterways. The island of Walcheren, deliberately flooded by the RAF bombing its perimeter dike, all but disappeared beneath the sea. The *New York Times* correspondent wrote that the water came in 'not as a kind of tidal wave that engulfed villages, but ... with the speed and deadly intent of a serpent'.

In 1945, the Netherlands was liberated and the Rijkswaterstaat had a new director-general to shoulder the task of rebuilding a shattered country's dikes, roads and bridges. The man appointed was van Veen's nemesis, Willem Harmsen. With persistence but little tact, van Veen presented him with a report on the state of the dikes in Zeeland. Without a word, Harmsen walked across his office, placed the file in a cabinet and locked it.[8]

Van Veen was not the only one with doubts about the state of the country's defences. One of his colleagues at the Rijkswaterstaat believed the Dutch approach to building dikes – including their most spectacular effort, the closure of the Zuiderzee – was fundamentally flawed.

Traditionally, Dutch experts had looked to the historically highest flood levels, then added another metre as a margin of safety. This principle had determined the 7.25-metre height of the Afsluitdijk. Pieter Wemelsfelder, a hydraulic engineer whose round spectacles and bouffant hair gave him the look of a startled owl, disagreed. Beginning in 1939, he developed an entirely new approach to determining the level of flood protection his country needed.

'The maximum stormflood, known by observation, is of little importance,' he wrote.[9] 'It may be just a quite arbitrary choice by nature out of the whole range of possibilities.' Across a short time-span, the 1953 disaster was Gulliver in Lilliput, Wemelsfelder argued, towering above other floods that had hit the Netherlands in living memory. But over the span of 1,000 years, it was almost certain – a 97 per cent chance – that the 1953 flood would be exceeded. It was still Gulliver, but now in the land of the giants. What mattered wasn't history; it was probability.

Wemelsfelder's fellow engineers were sceptical of measuring time in centuries, a span far outside a human lifetime. But he pointed out that while this was longer than a human life and even the life of an individual barrier or earthwork, the communities such barriers were meant to protect needed to ensure their survival for much longer – against the eternal threat from the sea.

And there was another factor: the question of what lay behind the flood defences. Wemelsfelder had been born in the south-western province of Zeeland, which was hardest-hit by the 1953 storm surge, but his thinking about floods was cool and rational. If, like the dike that was plugged with the boat *Twee Gebroeders*, the defence of Amsterdam and Rotterdam was at stake, then vast expense could be justified.

But if a barrier was built at great cost to defend a handful of isolated farmhouses, it was harder to make the case for the use of government funds. It was the same logic that had led Sir Hermann Bondi to push for the creation of the Thames Barrier, and one that seems hard to dispute now, but Wemelsfelder's work laid the foundation for this shift in thinking. What mattered, the engineer argued, was to take the two together: weighing up both the risk of severe floods and the cost in human lives and destruction of property if water overwhelmed the defences.

Bondi made this point very bluntly in his report on the Thames.[10] The United Kingdom can live without the oil refineries of the Essex coast, or even without Canvey Island, he wrote, 'but without central London it is a headless body'. In a world that has become more sceptical of elite opinion, Bondi's language can seem dated, mildly offensive even, but he confidently argues that the loss of a large number of 'highly qualified people' would impoverish the lives of everybody in the country. The approach developed by Wemelsfelder and championed by Bondi became the way that every society now judges risk.

By 1952, while the post-war reconstruction of the Netherlands was underway, van Veen feared the coming of a storm that could sweep all of this progress away. He was growing increasingly frustrated and decided to abandon all caution. Breaking the usual etiquette that requires civil servants to keep their views private, he arranged a trip to Zeeland with Herman Looman, an acquaintance who was a reporter for the magazine *Elsevier*. The country faced disaster, van Veen told the journalist, and 'people just don't get it'.[11]

The risk was not just to Zeeland. The engineer explained that the Schielands High Sea dike, on the Hollandse Ijssel, had many vulnerable spots. He told the journalist that 'millions live here below sea level, without realising what this means when disaster strikes'. Van Veen did not spare anyone's blushes; he criticised the system that split responsibility between the Rijkswaterstaat and local water boards, he blamed the Rijkswaterstaat for failing to intervene to maintain weakening dikes and he attacked his own managers for not facing up to the peril the country was in.

It was a career-ending interview. Or it would have been, had it been published. Looman's editor at *Elsevier* feared it was alarmist and, worse, that it would put readers off. The Dutch had suffered under the Nazis for five years, the editor reasoned. Telling them now that everyone was going to drown would be bad for business. He spiked Looman's story (although it resurfaced, 25 years later, when the journalist wrote a commemorative article about 1953).

Van Veen was disappointed, but not deterred. Harmsen had retired in 1951 and there was hope that a new director-general might listen. He prepared a new version of his plan to seal the Netherlands from storm surges, accompanied by maps of what the area would look like in future.

Building a series of giant barriers to close estuaries and inlets of the sea would create a shorter and stronger line of defence along the coast, van Veen reckoned, requiring less effort to maintain than hundreds of kilometres of dikes. But closing the estuaries of the south-west Netherlands was, he knew, a formidable engineering challenge, with strong tidal currents continually threatening to sweep construction material away.

The closing of the Zuiderzee offered a template, but that had been a single claw of the water-wolf and the bottom of that sea offered a sturdy mix of rock and clay on which to build a dike. In the south west, the water-wolf had many claws, slashing deep into the land, and the engineers would have to build on sand. The largest inlet, and the one least studied by water experts, was the Eastern Scheldt, which extended for 50 kilometres into Zeeland province.

The plan was dated 29 January 1953. Cassandra's final warning.

When disaster struck, even van Veen was shocked. 'As children, we had been told that we had a dragon living next door,' he wrote. 'And of course we knew it was there, but we had never ourselves seen and felt the quick, sudden, tiger-like stroke.'[12] After the flood, van Veen laid a book on his desk in the Rijkswaterstaat building and kept it permanently on display. It was a record of all the sea floods in the Netherlands, more than 100 in seven centuries.

This terror, he thought, must end. Alongside the 1,835 people who lost their lives, more than 72,000 had been evacuated. In a few places, the dikes had not just been breached but completely swept away, including the island of Goeree-Overflakee in the south west, where 24 kilometres of earthworks had been destroyed. The country was reeling.

Days after the disaster, on 18 February 1953, the Dutch government created the Delta Committee. The question the government asked them to solve was this: whether it was better to raise the dikes or find a way to close off the estuaries and sea inlets, as van Veen proposed. It was an idea that sounded simple, but that would test the limits of existing techniques and technology. Van Veen, his hour come at last, was appointed the committee's secretary.

The Delta Committee backed van Veen's proposal, allowing just two exceptions: the Rotterdam waterway and the Western Scheldt, the estuary between the southern Netherlands and Belgium, which was the shipping route to Antwerp. These waterways would need to stay open for commerce. Reinforcing the dikes here would provide greater safety.

They agreed that the most vulnerable spot in all of the Netherlands was the Hollandse IJssel river, where Captain Evegroen had held back the flood with his boat. The committee urged the government to begin with this river, building a barrier with movable gates that could be closed to seal the waterway if a storm surge threatened the centre of the country again.

The engineers ranked the works in order of difficulty; a barrier across the Hollandse IJssel was the most practical. The one that would strain their ingenuity most would be closing the Eastern Scheldt. It was wide and deep, with powerful tidal currents flooding and ebbing through it twice a day, shifting vast quantities of sand as they did so. At this estuary, and all along the coast, the threat of storms posed a constant risk of disruption to engineering work.

Van Veen and the other experts of the Delta Committee adopted Wemelsfelder's approach to the height of sea defences, looking at probability rather than history. The 1953 storm surge was a one-in-300-year event, which meant that in nine centuries, there were likely to be three storms of that scale. The committee decided that the level of protection for the most densely populated and economically valuable parts of the Netherlands – the heart and head of the country around Rotterdam, Amsterdam and The Hague – would need to be far higher. Flood defences should be able to cope with a storm up to the level of 'one in 10,000 years', the most exacting standard the world had ever seen. The rest of the country should be protected to a standard of 'one in 4,000 years'. The Delta Act, approved by the Dutch parliament in 1957, turned the committee's proposals into law.

Van Veen's book was republished, with a new chapter written by 'Dr Cassandra', outlining the 1953 flood and its consequences. In passing, this new chapter notes the planet's altering climate: 'It is to be expected that the sea will continue to rise … in consequence of the melting of the ice in Greenland and the Antarctic.' If the current generation does not face up to this shift, Dr Cassandra writes, their descendants will have to act. Closing the coast against the North Sea by sealing tidal inlets was, in van Veen's framing, the apex of centuries of progress. Where nature had branching rivers that discharged into curving estuaries, engineering would create straight lines. Van Veen's giant dams would reduce the length of the Dutch coastline by around 700 kilometres and turn the river delta, where saltwater flowed in and out with the tide, into freshwater lakes.

All resistance to van Veen melted away. The Rijkswaterstaat agreed to construct a machine, the Deltar, to test his idea of using electrical currents to simulate the tides in the Rhine, Meuse and Scheldt rivers. The apparatus, one of the earliest computers, occupied a room in one of their buildings. But van Veen's health was declining and, in 1957, he suffered another heart attack. The engineer attempted to carry on working, but the episode left him so incapacitated that he occasionally had to spend the whole day in bed. He succumbed to a final heart attack in 1959, as he was taking a train to The Hague a few weeks short of his 66th birthday. He was a prophet who would never see his promised land.

Van Veen's death marked a pinnacle of optimism in techno-logical ingenuity, the march of progress outlined in *Dredge, Drain, Reclaim*. The Dutch use the word *maakbaarheid* ('makeability') for the utopian idea that nature and society can be remade according to human wishes. But it was not just the Dutch who espoused it; the concept was popular in every advanced nation. Since ancient times, rulers had understood that, in times of abundance or scarcity, when there is too much water or too little, the power to control the flood, irri-gate the land and hold back the sea was also the power to shape people's lives. This had accelerated with the advent of capitalism and the Industrial Revolution, which allowed the deployment of money, labourers and machines on a larger scale than ever before.

The country that made greatest use of water engineering to shape its destiny was the US. In the early 20th century, the Colorado River, which tracks a path south west across the western states, brought regular and disastrous floods. The river rises in the Rocky Mountains, flows through great stretches of desert and discharges into the sea in the Gulf of California.

Taming its water was, American leaders understood, essential not just in order to contain the floods, but also to open up the arid west to farms and cities. Reclamation is the word used in other countries to mean turning swamp into farmland, but the task of the US Bureau of Reclamation, created in 1902, was to build the dams that would make the desert fruitful. 'Here is a land where life is written in water,' a Colorado poet wrote.

The bureau's engineers created the Hoover Dam, along with around 50 other barriers that bridle the flow of the Colorado River. Lake Mead, penned behind the 220-metre-high concrete cliff of the Hoover Dam, provides the drinking water that makes life in Las Vegas possible, as well as supplying Los Angeles and Phoenix. A bureau pamphlet from 1965 summed up its faith in technology: 'Man serves God. But Nature serves Man.'

Humanity's 20th-century dam-building spree has been an environmental disaster, disrupting the migration of fish, such as salmon, who battle their way upstream to lay eggs in the same stretch of river where they spawned themselves. Dams block the flow of sediment, which used to settle along the floodplains downstream, enriching the fertility of the soil.

By the late 1960s, Dutch engineers had begun constructing three artificial islands in the Eastern Scheldt in preparation for building a dam that would seal the estuary from the

North Sea. It was to be the last and most complex of the Delta works, but there was growing concern about its potential environmental consequences. Rachel Carson's book, *Silent Spring*, published in 1962 and translated into Dutch soon after, was an influence, popularising the idea that humanity could alter and destroy parts of the natural world, and that this harm would in turn rebound on our species.

Carson was a marine biologist and, before *Silent Spring*, she had written a popular trilogy about the sea. Her writing sought to connect people with the beauty and rhythms of earth and sea, and the power this had to inspire and console people, contrasting it with the concrete and steel with which humanity was insulating itself from natural reality.

One of the leaders of the environmental protest in the Netherlands was Jan Loeff, a naval architect and passionate sailor who settled in a waterfront house in Veere, a picturesque town of cobbled streets, merchant houses and a marina, on the island of Walcheren. As well as sailing his own yacht, Loeff was editor of a boating magazine *Waterkampioen*, 'Water Champion'.

Loeff made the case as matter-of-factly as possible.[13] When the tide ebbed in the Eastern Scheldt estuary, the mudflats became a hunting ground for wading birds like the oyster catcher. As it flooded again, ducks dived. Migrating geese wintered there, while mussels and oysters filter-fed in the mix of salt and freshwater. Grey seals lounged on seabanks and a porpoise's dorsal fin occasionally broke the water's surface. The river delta and estuaries were a living system of shifting water and sand, and the presence of the sea meant that summer temperatures were cooler and winters milder than the surrounding land.

The sea inlets of the south-western coast were not simply a threat to human existence at times of exceptional storms but also, all year round, places of natural variety and beauty. And, because of the presence of oyster and mussel beds, they had economic value too. Slicing across the Eastern Scheldt with a barrier of concrete and steel would disrupt this system, just as dams blockaded living rivers. When the sea was cut off and the water turned fresh, many species would disappear. The living coast was not just a metaphor for the movement of current and silt, but a biological fact too. This life provided its own form of flood protection, if it was left undisturbed; plants' root systems slowed the velocity of floodwater and protected the beach from erosion.

The Delta Works themselves provided an illustration of the risk to nature. Grevelingen, an estuary just north of the Eastern Scheldt, was sealed off from the sea in 1971. The saltiness of the water was quickly diluted by rain, while the sand flats that had been submerged at high tide and exposed at low tide were left permanently dry. With these flats no longer washed by tides and wriggling with worms and shellfish, the numbers of wading birds plunged. Out of ten species of lobster and crab, only two survived the abrupt change. Fish that migrated to spawn found their path blocked. Van Veen's design was taming the unruly sea, but choking the life out of it.

Alongside the fear of the 'water-wolf', there was another tradition in Dutch history: one of living with the water. The environmental historian Petra van Dam has described this as an 'amphibious culture', in which churches were sometimes built explicitly as refuges both for people and their cattle, while farmers kept boats both to take their goods to market and to escape when the dikes were breached.[14]

Vast reclamation projects like the Afsluitdijk enclosing dam and flood defences like the Delta Works spelled the end of this culture. Cities expanded into low-lying areas, including the bottoms of drained lakes. The Haarlemmermeer polder, reclaimed from a lake in the 19th century, is now the home of Schiphol Airport. Van Veen's fortifications were the crowning glory of a culture that regarded the sea as a permanent enemy, but now sailors like Loeff joined forces with fishermen and environmental campaigners to remind their country of another way to see the coast.

Loeff called for an inquiry, based on the latest scientific developments, to consider alternatives for the protection of the coast. After the 1953 floods, the Delta Committee had acknowledged that closure of the Eastern Scheldt would leave it 'utterly transformed', with the loss of many animal and plant species, but decided that safety overruled the loss to nature.

Nearly two decades later, with growing awareness of the environmental consequences of dams, the project became the focus of sustained opposition. Ahead of elections in 1972, parties on the left promised to revisit the decision. The Rand corporation, a US think tank, was called in to advise and they presented their findings in a briefing at the Rijkswaterstaat's headquarters in April 1976.

The Rand study, conducted jointly with the Rijkswaterstaat's experts, did not make a single recommendation, but instead looked at three possibilities.[15] The Eastern Scheldt could be closed completely with a dam. This would be worst for the environment. Or it could be kept open and the dikes around it strengthened. This would be worst for flood security. The third option was to build a storm surge barrier with gates that would allow the tide to ebb and flow in normal weather conditions, but could be lowered when a surge threatened to whip the sea inland. This would be the most expensive.

The centre-left government that formed after the 1972 election backed the creation of a storm surge barrier in the Eastern Scheldt. It was a compromise, preserving nature while protecting people against floods. The only remaining question was how it would be accomplished. The original plan had been to drop rocks into the Eastern Scheldt from the gondola of a cable car running across the estuary, the same method that had been used to close the Grevelingen estuary from the sea. But constructing a barrier was a more complex affair than a dam.

Instead, across the 8 kilometre-wide channel that was the chosen spot for the barrier, 65 giant concrete pillars, each one as tall as a parish church, were planted in the fine sand of the estuary bed.

Between the concrete monoliths, heavy steel gates were installed. There are 62 of these gigantic steel slabs and the heavier ones weigh 480 tonnes, about three times as much as the Statue of Liberty.

A specially designed ship – the *Ostrea*, a floating platform with a giant industrial crane on its deck – lifted the barrier's concrete pillars into place. This task required a combination of immense power and precision, as the pillars needed to be placed exactly on their foundations, spots where the seabed had been compacted so the columns would not sink or shift in the fast-flowing current. The maximum room for error when placing each of the pillars was a single centimetre.

The Dutch were so delighted with their achievement that they put a plaque at the end of the barrier. It read: 'Here the tide is ruled by the moon, the wind, and us.' The barrier was completed in 1986, with Queen Beatrix declaring that 'nature is under control but not disturbed'.

The last part of the Delta Works to be completed was the Maeslant barrier, built between 1991 and 1997 to protect the waterway linking Rotterdam with the North Sea. The barrier has two giant triangular steel arms, one on each side of the waterway. Each of the arms widens out into a hollow, curved gate more than 200 metres across. When a storm surge threatens Rotterdam, the arms swing out and bring these monumental gates together.

Inlets in the machinery are opened to let the gates fill with water, and they slowly sink down to the bottom of the waterway. From above, the triangular steel arms look like two Eiffel Towers lying flat on the ground. Stand next to it on the ground and the lattice of beams resembles the steel skeleton of a theme park rollercoaster.

Van Veen's barriers had armoured the Dutch coastline and, once the last gap in this armour was closed, it was easy to overlook the country's enduring vulnerability. The densely populated crescent of land around Amsterdam and Rotterdam is barely above sea level – and parts of Rotterdam are more than 6 metres below the water line – while the great rivers that wind through the Netherlands make it a storm drain for western Europe.

The pride in Queen Beatrix's speech may have been justifiable, but it also smacked of over-confidence in the human ability to transcend natural forces. This hubris would be punished less than a decade later.

CHAPTER 6

YOU CAN'T FIGHT WATER

The floods of 1953 came as a shock partly because they had followed centuries in which the British and Dutch had engineered their rivers and coastlines. This was true in nearly every region of the two countries, but one of the deepest changes had taken place in the counties of eastern England that were hardest hit by the North Sea disaster.

Eastern England was once a vast landscape of wild wetland: the Fens, a treacherous territory of marsh and forest linked by shifting channels of water with the occasional napkin of firmer ground. For the men who knew their way round it, by boat in summer and on skates in winter, it yielded a rich harvest of fish and eels, reeds to thatch roofs and, most valuable of all, the lush grass that sprang up to feed cattle after the great rivers flooded.

Acquisitive kings and noblemen had eyed the land for generations, knowing that draining the fens would uncover one of the richest soils in England, but lacking the means to undertake such a vast project. The idea was spoofed in a play by Shakespeare's contemporary Ben Jonson, *The Devil's Ass*, in which the gullible gentleman Fabian Fitzdottrel is hood-winked with a get-rich-quick scheme to make him Duke of the Drowned Lands.

The swindler Merecraft makes it clear that the draining of the fenland will benefit the Crown and wealthy backers, while ordinary citizens will be brought in as small investors, after which the big players will 'blow 'em off again/Like so many dead flies'. The mockery was a little uncomfortable for King James I, who had just granted rights to drain and enclose fenland owned by the Crown, and he asked the playwright to remove the offending lines.

Within a few years of *The Devil's Ass* first being performed in 1616, a real-life Merecraft promised to turn the play's satire into fact. There's no monument to Cornelius Vermuyden, a Dutch engineer who arrived in England around 1621, when he would have been in his early thirties.[1] There's not even, with any certainty, a surviving portrait of him. Not much is known about his youth, beyond the fact that he came from a family of water engineers from the southern Netherlands, but he arrived in England with swagger and influential connections, and would go on to leave an enduring mark on the geography of the country.

Vermuyden's work would create the most productive farm-land in Britain. To this day, the fertile earth of East Anglia produces a third of England's potato crop and nearly a third of its wheat. His engineering would also mark the beginning of a new relationship between people and the natural world. While humans have sought to exploit and control the flow of water from the dawn of civilisation, the growth of finance and trade in the early 17th century meant that grand projects, which would once have been fantasy, could now muster sufficient money and labour to reshape earth and water.

The Dutch had already turned marsh into farmland, but in England, a country three times bigger than the Netherlands, Vermuyden's transformation took place on a vast scale. Under King James, his successor Charles I and the Commonwealth that emerged after the English Civil War, the wild fenland would be domesticated. The draining of the Great Level – a 300,000 acre expanse of East Anglian fens – ranks as one of the marvels of 17th-century engineering. One of the Environment Agency officials now tasked with overseeing Vermuyden's legacy likens it to the creation of the Palm Jumeirah, the artificial islands off the coast of Dubai.[2] It was a moment when the English looked at their landscape and decided they would transform it completely.

Vermuyden's name lives on along England's east coast. After severe flooding across the UK in the winter of 2013, Andrew Percy, an MP for a constituency on the Humber estuary, told the House of Commons: 'We know we live in an area that was drained by Cornelius Vermuyden in the 17th century. It is former marshland.' The MP's point was that his community knew their history. As he called for funding to strengthen flood defences, Percy said: 'We know the risks we face. That does not mean we should be written off.'

Like many innovators, Vermuyden was combative. His first project in England – to repair breaches in the banks holding back the Thames at Dagenham in Essex – underlined that. The engineer had been introduced at the royal court through

129

his brother-in-law, a Dutch diplomat, helping him secure the commission. Instead of simply repairing the existing breaches, Vermuyden tore down the banks and rebuilt them at a distance from where the river flowed, a move that, while giving the water space to spread out when the river was swollen by rainwater, sacrificed valuable farmland.

In common with many overseers of public works before and since, he also appears to have significantly underestimated how much this would cost, leaving workmen complaining that they had not been paid. The local officials responsible for drainage of the land said he had left it in a worse state than before and refused to pay him. The king's advisers urged compromise, but Vermuyden and the local landowners were unable to agree terms. The Dutchman's response was that he would pay the workmen – just as soon as the officials paid him.

When King Charles I came to the throne in August 1625, Vermuyden was granted a portion of the reclaimed land by the Thames as compensation. The episode signalled that he was in royal favour. Throughout his career, he was both an engineer and a speculator, prospering as he reshaped the land.

In 1623, Vermuyden was contracted to build sea walls encircling the marshes of Canvey Island. This was good grazing land for sheep, but local landowners wanted to make a permanent settlement here. The oak stakes that provided the foundations of these coastal walls are still visible along the shoreline, though the chalk and clay that filled them in have been obliterated by the sea. He worked in partnership with a London-based Dutch merchant, who financed the venture.

Three years later, King Charles, by now firmly established as his royal patron, authorised Vermuyden to drain Hatfield Chase, an expanse of marshland on the border of Yorkshire and Lincolnshire. Again, Dutch engineering was combined with Dutch finance, now from a consortium of investors.

Here, Vermuyden shifted winding rivers from their natural courses by excavating artificial channels running in straight lines to make the water flow faster. But his rerouting of the River Don, which swelled the volume of water flowing down it, brought flooding to villages that had previously been dry. It was, at the time, the largest land reclamation that had ever taken place in England. And it was financed and managed almost entirely by foreigners (though Vermuyden had been naturalised in 1624), while Dutch labourers were brought in to carry out the work, rather like the Chinese-run and -financed road and railway building projects in Africa today.

In letters from the time, the work that was done to reshape England's eastern flank was described as 'improving' the land. And, of course, it did. A straightened river is easier to navigate than a winding one, while a drained marsh can be turned into a field of wheat. But it also rendered the terrain more fragile, setting up a permanent competition between technology and nature. The land taken from rivers and the sea across the country could be swallowed up again. And now it was no longer marsh that served as grazing land for sheep, but people's lives that could be lost, and valuable homes and crops destroyed.

And the ordinary people who lived in the region did not regard it as an improvement. Hatfield Chase itself was a royal hunting ground, but it was bordered by a ridge of higher ground called the Isle of Axholme ('isle' because it was less likely to be flooded than the surrounding marsh rather than because it was a true island). Vermuyden's grand design excluded villagers who used the marshy land to graze their livestock, to fish and trap wildfowl, and to gather peat and turf for fuel. The draining threatened all of this and people retaliated by tearing down Vermuyden's embankments. The confrontation escalated into violence in 1628, when a local man was killed by one of Vermuyden's Dutch associates.

Sir Robert Heath, King Charles's attorney general, intervened on Vermuyden's side. The rebellious villagers were ordered not just to cease attacking the works, but to help build them. According to a letter from one of the Dutch financiers, a royal proclamation was read out in the village accompanied by 'threats of fire and vengeance'. It's a scene that might be expected on the oppressed fringes of an empire rather than in the heart of a European nation: a cash-strapped monarch, allied with foreign capital and expertise, 'improving' his subjects' land against their wishes and threatening them with violence if they failed to comply.

But now, as in Dagenham a few years earlier, things began to go wrong. Vermuyden's work to 'improve' Hatfield Chase faced resistance from nature itself. The Dutch engineer created 'washes' that gave the rivers room to spread when they flooded, but there was a flaw in his design. Vermuyden identified the Don, which flows down from the Peak District and winds through Hatfield Chase, as the principal culprit for the land's dampness, and his scheme blocked off two of the river channels that emptied into the River Trent in order to divert its entire flow northwards into a branch that emptied into the River Aire.

The engineering of the Don worked perfectly for the land that Vermuyden wanted to drain. It worked much less perfectly for people in a cluster of Yorkshire villages north of the Don. The volume of water was too great for a single channel and these villages were now exposed to flooding on a scale they had never previously encountered. Local officials ordered Vermuyden to cut a new channel to provide flood relief and the backers of his scheme had to cover the costs, a vast additional expenditure that tipped the project into financial chaos.

The relief channel for the Don gave the river a new outlet to the Ouse, which flows out to the Humber Estuary. Vermuyden was asked to pay a share of the costs of this scheme. When he failed to contribute, he was imprisoned by northern lords who seemed unimpressed by the fact he was a royal favourite. The Don's unwillingness to follow human direction persisted, as the water abandoned Vermuyden's original channel to the Aire and flowed entirely down the flood relief channel. The stretch of the Don flowing down this relief channel is still known as the Dutch River.

The civil war that broke out in England in 1642 paused Vermuyden's ambitions. The villagers neighbouring the Hatfield Level, who had rebelled against the drainage of the marsh, took advantage of the disorder to wreck the work again, at one point deliberately opening floodgates to inundate land that had been drained. But while there was plenty that divided the king and Parliament, the taming and improving of the country's 'wastes' – as swampland was often called – was a common desire.

Vermuyden's reputation had suffered after Hatfield Chase, but despite everything that had happened, he was still regarded as the country's foremost expert in land reclamation. In 1649, when investors led by William Russell, the earl of Bedford, proposed to drain an expanse of East Anglian fens called the Great Level, they turned to Vermuyden – albeit with some understandable reluctance.

Eric Ash, a historian and the author of *The Draining of the Fens*, said: 'They know he's problematic, but there's no one else available with the skills, the plan, the confidence. It's a year of negotiation while they try to find someone else.' The investors eventually took Vermuyden on and then found him impossible to control. 'They don't like the fact he won't give updates. They fear he's being wasteful with their money,' Ash said. 'His argument is – you don't understand what I'm spending money on. I know more than you do.'

Among the backers of the project was Sir John Popham, the judge who presided over the trial of Guy Fawkes and the other Gunpowder Plot conspirators. In common with many other investors in Fen drainage, Popham also supported projects to settle English farmers in Ireland and establish a colony in North America – a settlement, the Popham Colony, was named after him in what is now Maine. This was not just coincidence; the drive to colonise and exploit the North American 'wilderness' and the reshaping of the Fens shared a disdainful attitude to the territory and the people who already lived in it.

In East Anglia, Vermuyden and his investors faced the same engineering challenge as at Hatfield Chase but on a far larger scale, with six rivers winding slowly through low-lying terrain. Vermuyden's proposed solution was the same, cutting a straighter, steeper channel for the principal river – here, the Great Ouse – that would accelerate its journey out to the North Sea. He cut two straight channels running in parallel to carry the flow of the Great Ouse, the Old Bedford River and the Hundred Foot Drain (called that because it was 100 feet between the embankments on either side of the channel). He left space between the two – a 'wash' – to allow room for the water to spread in winter. This zone is now called the Ouse Washes, a meadowland in summer that is submerged in winter; it's a haven for wild birds. As at Hatfield Chase, though, the work faced sabotage from locals who smashed sluice gates and destroyed embankments.

By 1652, the year after the civil war ended, thousands of acres of land had been drained and the scheme appeared a success. But, once again, his actions wrecked a fragile balance in nature. The land he drained was peat, a combination of living vegetation and the slowly decomposing remains of plants and the microbes that break those plants down. As well as being host to a wide range of plant and animal species, peat is a significant store of carbon and the destruction of peatland for agriculture around the world has destroyed one of the planet's primary means of locking away a greenhouse gas.

When the peat soil of East Anglia's fens was drained and exposed to the air, it began to dry out and decompose. As it did, the drained land subsided and flooded again. By the end of the 17th century, large areas of the reclaimed land were regularly underwater, regarded just as much as 'waste' as they were before Vermuyden had started.

The Dutch engineer's work across eastern England was, by ecological standards, a disaster. The landscape was once made up of winding rivers and wetlands with banks of reeds and willow trees, providing sheltered spawning grounds for fish, as well as food and hiding places for migratory birds, and space to thrive for otters, water voles and deer. The channels he created were faster and less sinuous, with much less shelter for fish and birds to rear their young. The peat, once uncovered, dried out and was eroded by the wind.

However, his design set a template for the future. Further engineering work continued over the next few centuries, adding more channels to carry water away and sluice gates to control the flow – gates that are opened when the water is high, speeding its flow, or closed when water is too low. The fields had now sunk below the level of the rivers that drained them, so windmills were installed to pump water out. These windpumps were replaced by coal-fired steam pumps, then diesel ones and now electric pumps. The abundant fields of East Anglia are kept dry by the constant vigilance of machines.

The draining of the Fens was part of a broader reshaping of waterways, creating canals that speeded the shipment of coal from inland mines as well as delivering finished goods from potteries and steelworks. The Fens once provided copious eels and waterfowl to feed its own people. Once drained, the land yielded potatoes and wheat. As Britain industrialised, the country imported vast quantities of food, but these home-grown harvests helped feed the factory workers who crowded into Birmingham and Manchester.

It marked a philosophical shift too. The investors backing the draining of the Fens hired William Dugdale, a scholar and court official, to write an account of their project. In a book published in 1662, *The History of Imbanking and Drayning*, Dugdale sets it in the grandest possible context. Rather than accepting the world as God made it, Dugdale casts the Creator as the first engineer, gathering the waters together and letting dry land appear at the beginning of Genesis, and again after the Biblical flood.

The writer, who specialised in the study of ancient history, then tours the drainage projects of Egypt, Babylon, Greece and Rome, putting Vermuyden's exploits on a pedestal along-side the feats of antiquity. It's a testament to the endurance of the British elite that one of William Dugdale's descendants, Sir Bill Dugdale, was chairman of Severn Trent water author-ity in the early 1980s, at a time when the water authority was engaged in draining wetlands in Nottinghamshire.

While a modern ecological view might equate Vermuyden's work with taking a chainsaw to rainforest, the draining of the Fens helped establish the principle that a landscape could be 'improved' through technology, being turned from a waste that benefitted a handful of local people into productive and privately owned land that benefitted the whole country. This would become a regular feature of Britain's imperial expansion. 'What the British do in the Fens, they try to do in Ireland, they try to do in North and South Carolina,' says Eric Ash.

In the 1840s, the British built what remains the world's biggest irrigation system – not in their own country but in the province of Sindh, on the Indian subcontinent. The system of canals diverted the flow of the Indus River, which streams down from the Himalayas and the Hindu Kush to the Arabian Sea. The Indus Valley became valuable farmland, growing indigo for export.

After independence, when Sindh became part of Pakistan, the transformation of the Indus continued, with many more barriers added to alter the river's natural flow. Landowners became immensely wealthy, but so much water was diverted that, in many years, the Indus did not reach the sea. The taming of nature that began with Vermuyden in 17th-century England would go on to have deadly consequences around the world.

The eastern coast of England is now protected by a long line of sea defences, rebuilt and strengthened after the 1953 flood. Towns and villages inland nestle in an artificial landscape where water must constantly be pumped out to sea. At its lowest, the Fens are up to 2 metres below sea level. 'If the sea level rises by a metre, we get to a point where the picture is quite scary,' that Environment Agency official said.[3] 'The challenge is to get the water out to sea, out to a sea that's higher. If the coastal defences breach, the North Sea doesn't stop. It just keeps coming.'

The work that Vermuyden began and his successors perfected appeared to have tamed nature, but water would not always flow as humanity intended. Instead, the reshaping of rivers and coastline created new danger.

By the standards of a European winter, the start of 1995 had been unusually warm. Heavy rain combined with rapid snow-melt to send millions of litres of water surging down the Rhine, sweeping across Germany until it reached the south-ern Netherlands. Ed d'Hondt, mayor of the largest city in the region, Nijmegen, feared that the dikes that held back river water were on the verge of collapse.

Nijmegen lies between the Waal – as the main branch of the Rhine is known in Dutch – and the Maas, which flows down into the Netherlands from France, where it's called the Meuse. Behind the dikes lay polders that fill up like bathtubs if the earthen ramparts crumble. As the rivers swelled over a few days at the end of January and the beginning of February 1995, Dutch troops were sent out in a desperate effort to reinforce the dikes with sandbags. Despite this reinforcement, d'Hondt still feared that the escalating pressure from the water would shatter the defences.

He ordered the evacuation of the entire region, home to around a quarter of a million people. It was the end of January 1995, exactly 42 years after the 1953 flood. The television news that night began apocalyptically: 'The Netherlands tonight faces a total emergency,' declared the newsreader calmly, as TV screens showed pictures of people, clutching a few possessions, boarding buses to be evacuated to safety.

Carina Verbeek knew something was different when her father came home from work at lunchtime, on a rainy day in early February.[4] But the eight-year-old was excited rather than scared. She helped her parents carry furniture upstairs, hoping it would stay dry, and then, clutching Oof, her cuddly toy elephant, got in the car as the family prepared to drive north with their cats Mick and Muis.

Their village, around 50 kilometres west of Nijmegen, lay by the River Lek, a branch of the Rhine, and was threatened by dangerously high water levels. The Rhine floods regularly, but the combination of warm weather accelerating the melting of snow and heavy rainfall was exceptional. At the German border, where the Rhine usually rolls into the Netherlands between 7 and 12 metres high, the river had reached 16 metres. The Verbeek family headed to Carina's aunt's house, the traffic inching along narrow roads as everyone headed away from the rising water.

A stretch of dikes around the village of Ochten, between two branches of the Rhine, was the weakest point in the whole country's flood defences.[5] They were crumbling dramatically under the weight of water. There was, d'Hondt told journalists, about a 50/50 chance that the dikes would collapse. The Betuwe region, where the Verbeek family lived, is a terrain of lush green fields and rivers fringed with willow trees, famous for its apples, pears and cherries. It would all be underwater if the Ochten dikes collapsed.

With water seeping through the dike, the villagers in Ochten were evacuated. As the Verbeeks had done, families put their furniture up on the first floor, grabbed their children and pets, and fled. Across the region, roads were flooded. In some places, the water levels rose parallel with the rooftops of houses. Police patrolled in motorboats to ensure everyone had left, and to deter looters.

And then, in early February, the rains in Germany and the Netherlands eased and the water levels began to subside. As people returned to their homes from the biggest evacuation since the floods of 1953, the vast crisis provoked a reckoning. The rivers that made the region so fertile, and such an idyllic place to live, had turned into menacing expanses of brown water. The Dutch had bolted the front door after 1953, but now found water seeping in the back.

Wim Kok, the prime minister, voiced the instinctive reaction: the government would match the Delta Plan that held back the North Sea with defences against the rivers. It was just two years since the last time the rivers had flooded, in the winter of 1993, when the Rhine burst its banks in Cologne and Bonn while high water in the Meuse forced the evacuation of thousands in the south-eastern Netherlands.

The danger from the rivers, the prime minister said, appeared even more threatening than the sea. But others asked whether snow and rain alone were to blame for the crisis.

The upper reaches of the Rhine once meandered through meadow and pine forest in southern Germany, splitting into braids that curled around islands of rock and silt, regularly flooding and receding through boggy wetlands glittering with dragonflies. Then, under the direction of Johann Gottfried Tulla, a German civil engineer, it was straightened and shortened.

Tulla's work in the 19th century merged the upper Rhine's braids into a single channel, digging ditches that slashed straight lines through its wandering loops and fortifying its banks to confine the river in its new course. The wide floodplain was narrowed and its marshes 'reclaimed' for farming.

'No river needs more than one bed' was Tulla's law. This taming of the river continued under his successors, creating a single, faster path for river traffic. The romantic Rhine was celebrated by German patriots and admired by tourists on river cruises as aristocrats restored the gloomy medieval ruins on its banks, but it was made safe by removing its wild edges. A wetland forest in Plittersdorf in southern Germany, close to the French border, is one of the few survivors of the Rhine's past; a soggy landscape of water-loving trees that hops and crawls with amphibious and insect life.

The loss of the upper Rhine's natural floodplains led to higher flood waves downstream. Guarding against the risk of snowmelt in spring, cities like Freiburg on the upper Rhine are built on higher ground. Downstream, Cologne straddles the river, its builders trusting that the highest stretches of the river would absorb much of the melting Alpine snow.

By the mid-1980s, a group of Dutch landscape architects and environmentalists were prepared to break with the past. They proposed a plan to remove 'summer dikes' on the river banks – these were lower and closer to the river than the high winter dikes – and lower the floodplain by excavation.[6] Riverside forests would be allowed to grow back. Wetland birds, like the black stork, would return. Their idea, Plan Stork, gave the river space to breathe and let wild vegetation slow the movement of water rather than constantly raising the level of the dikes.

Closing the estuaries to the North Sea under the Delta Plan had been an environmental disaster. The tidal inlets, lapped twice daily by saltwater, turned into stagnant lakes where toxic algae bloomed, nourished by fertiliser washed off farmers' fields, while the migration of fish was blocked. Oxygen levels in the depths plunged due to the water no longer mixing as it did before, killing off the molluscs and worms on which other animal life depended.

The risk was not confined to the natural world. The river floods were a warning that building higher, stronger ramparts against the water might, in time, lead to a greater disaster when those dikes failed. And building higher dikes meant making them broader too, requiring the demolition of the houses next to them, which made it increasingly unpopular.

Dick de Bruin, one of the authors of Plan Stork, was an official at the Rijkswaterstaat who had an unexpected opportunity to make his case when he took a boat trip along the Waal with a government minister in 1989. The minister, Neelie Kroes, had a reputation for toughness and resolute decision-making; her nickname was 'Nickel Neelie'. Soon after he explained his idea, their boat passed a Rijkswaterstaat construction gang on the bank, extracting sand for a building project. 'If I understand you correctly, this isn't what we should be doing,' she said. De Bruin agreed. Left undisturbed, the sandy bank formed natural protection for the land behind it. Kroes picked up her phone and made a call on the spot to have the work stopped, batting aside protests from an indignant official on the other end. 'Get started,' she told de Bruin, telling him the small pocket of land on the south bank of the Waal was his to experiment with.

One of those whom de Bruin turned to was a young ecologist, Wouter Helmer, who had grown up in Nijmegen where he'd been a nature-loving boy, spending every spare moment of his childhood cycling out into the surrounding countryside with his brother Jeroen, trekking through forest and swamps, listening for the call of songbirds, cataloguing bugs and butterflies, and creating safe passages across busy roads for toads and salamanders. Wouter was the storytelling brother and Jeroen the artist, capturing insects and amphibians with a few deft lines of his pencil.

Helmer studied ecology in the 1980s when the environmental movement seemed deeply pessimistic, preoccupied by acid rain and the havoc that humanity seemed to be inflicting on nature. His own view, shaped by the lyricism of his childhood excursions, was admiration at nature's resilience. Animals and plants seemed to thrive in every gap that made itself available. Plan Stork's marriage of nature restoration and flood defence struck him as a story of hope.

The project to restore nature to the River Waal began with the 2.5-hectare strip of land on the riverbank offered by Neelie Kroes. The team's first act was to release three Koniks, a tough breed of pony. The Koniks, a stallion and two mares, would do their bit through grazing and by rolling in sand to clean their coats. Their cropping of plants helped create a mosaic of vegetation, while the sand-baths opened up bare ground that made space for insects to nest.

The Rio Earth Summit, a UN conference devoted to the natural world in 1992, strengthened their cause in an unexpected way. Conservationists from developing countries, pressured to protect their rainforests and coral reefs, turned the tables on the Europeans who had ravaged so much of their own nature. Chastened, the Dutch branch of the World Wildlife Fund threw its support behind Dick de Bruin's experiment. A video of the Koniks, galloping along the riverbank, helped swell public donations and allowed the team to buy neighbouring parcels of land.

Using historical maps, Helmer and his colleagues found blocked channels where the Waal had once run. Opening them up again would create new havens for wildlife, places for midge larvae to grow and create food for fish.

But although the minister had given the project her backing, and de Bruin himself was a Rijkswaterstaat official, other civil servants now frustrated the plan. The fear was that if the restoration let the river spread out into dozens of channels, the main watercourse would get shallower and clogged with sediment, and might become impassable for shipping. 'Now you're going too far,' Helmer and his colleagues were told.

And then the floods of 1995 hit. Friends whose riverside home was threatened by the rising water came to stay at Helmer's house in Nijmegen. Standing at the top of the dike by the Waal, he watched the swollen river rage past, just a few centimetres below the top of the earthworks. Behind him, thousands of houses clustered in a polder more than 6 metres below the Waal. Helmer and his colleagues rowed out to find the Koniks huddled on higher ground in a flooded swathe of the river bank, hungry but with their hooves still dry.

When the water receded, Helmer trekked along the floodplain and saw that the high water had transformed the landscape, leaving deposits of sand and clay in some places and carving out new gullies in others. There was life too; among the flotsam were the seeds of plant species that sprouted in the wet soil. Slicing open mushrooms washed downstream, he found beetles, ants and other insects hitching a ride inside. 'They're using those mushrooms like Noah's Ark,' he marvelled.

The near-disaster tipped the balance for the government. Plan Stork became official policy, under the name 'Room for the River'. At more than 30 locations across the Netherlands, work was carried out to reshape the relationship between land and river. Summer dikes were removed and new channels were dug to allow rivers to spread. The work was coordinated, the understanding being that the rivers were a system. The idea that went all the way back to the earliest days of Dutch history – that the sea can be restrained, marshes drained and rivers tamed – gave way to something new: not just accommodating nature but using natural forces to solve the ancient problem of floods.

Defying those who believed the rivers too polluted to harbour much life, forests grew in the floodplains, willow and poplar nearer the water, with oak and ash further up the banks. These created natural wind-breaks and blunted the impact of floods. It was a reversal of centuries of work to hold back the water, making an ally of nature rather than restraining it with concrete.

One of the biggest transformations to make more room for water took place at Nijmegen, where the River Waal narrows and bends east of the city. Nijmegen is the oldest city in the Netherlands, dating back to the Roman settlement of Noviomagus, and one that has prospered on river trade, with cobbled streets and towering brick buildings nestled close together in its rebuilt medieval market square. Facing it on the northern shore of the Waal is the village of Lent and this is where government planners turned to reshape the river. But dozens of families lived on the floodplain in Lent and they refused to move. It took more than a decade of bargaining before the householders accepted compensation and agreed to relocate.

Fifty houses were bulldozed to make space for the Waal and the riverside dike was demolished and moved 350 metres inland. A second 4-kilometre-long channel was excavated in the space that had been cleared, so the river now splits in two. Part of the old riverbank survives as a small grass-covered island in between the two channels.

A new bridge was built across the Waal, the riverfront promenade was renovated and the grass-covered fragment of the old riverbank became a park. The 'Spiegelwaal' – or 'mirror Waal', as the second channel is now known – is popular with outdoor swimmers. Building higher dikes or flood walls frequently cuts a city off from the water, which becomes invisible behind ramparts of earth or concrete. This has happened in London, where the river was once shallow and flanked by gravel beaches, allowing people to fish or go boating, but has steadily been surrounded by high embankments. In Nijmegen, the reconstruction did not just enhance flood protection; it meant the city turned to embrace the river.[7]

The work happened at a time when agriculture was growing more intensive, requiring fewer but larger farms. Smaller farmers sold up and bigger ones consolidated, which meant that strips of land became available to give back to nature. But another trend posed a threat. Cities were sprawling outwards and the river deltas – flat and well-connected – were always the most popular place to put new homes, sometimes with a view of the water. But natural defences require space. There was a narrow window in time to finish the task before the delta was engulfed with concrete.

The work of giving the rivers more space was largely completed by 2018 and, fewer than three years later, the rivers put the country's new defences to the test. Violent summer storms led to heavy rainfall that swelled waterways in western Europe in July 2021, sending floodwater tearing through towns and cities. Houses were crushed by the floods, streets transformed into surging streams and cars were swept away. Nearly 200 were killed in western Germany, with most of the casualties concentrated in the Ahr Valley, a wine-making region of steep, wooded slopes, where a river that was usually low and gentle turned into a wall of water.

The Meuse river, which flows down from the high plains of Langres in the Champagne region of France through Belgium and into the Netherlands, reached record-breaking levels as it crossed the Dutch border. But as the river swept on into the southern Netherlands that July, something was different. The Meuse spread into a new floodplain, wider and lower than before, absorbing some of its force.

When it reached the Netherlands, the Meuse was still high and powerful enough to breach a dike, which was rapidly shored up by soldiers with sandbags. Thousands of people were evacuated as a precaution and the centre of the town of Valkenburg, on the banks of the river, was flooded. But there were no casualties; Dutch towns and cities escaped much of the destruction inflicted on neighbouring countries.

More exposed to flooding from sea and river than any other nation in Europe, the Dutch had taken a series of imaginative leaps through their history that transformed their relationship with the water. Vermuyden and the engineers of the 17th century turned fenland into fields of wheat, understanding nature well enough to reshape the rivers but making the natural world serve man. Van Veen had made his country safe from the sea, although the Delta Works disrupted a living coastline.

From Rhine barges shipping coal to the tankers of Gulf oil and Texan liquefied gas that now flow through Rotterdam, man-made waterways, deep and straight, became the arteries that conveyed fossil fuels through Europe. The coastal defences were built of concrete and steel, requiring vast quantities of carbon to create, while the engines that pumped excess water out of the polders were powered by coal and then diesel. For more than a century, the Dutch thrived on planet-heating fossil fuels and turned to those same fuels to protect them from the consequences of a warming world.

The imaginative shift of the 1990s, creating wider and greener rivers, was the first sign of a move beyond that tragic bargain.

CHAPTER 7

DISASTER IN THE INDUS VALLEY

The south-west monsoon had blown in from the Indian Ocean in late June, bringing the rain that feeds Pakistan's thirsty cotton and sugarcane fields.

But this was not the life-giving gift of nature. That August, the monsoon poured down week after week, many times the usual amount of rain, which combined with unusually large volumes of water released by melting glaciers in the mountains of northern Pakistan.

When Sherry Rehman, the country's climate change minister, flew above the countryside in an army helicopter, she saw districts that 'looked like they were part of the ocean'. Instead of cotton's white blossoms, the landscape below was a sea of putrid green water.[1] In some places, helicopters were unable to find dry land on which to drop aid parcels and for the first time, Pakistan's navy was deployed inland to reach the least accessible areas.

Around a tenth of the country was submerged, claiming the lives of more than 1,700 people, and damaging or destroying more than a million homes. Much of the country's most fertile agricultural land was inundated.

Following a request from Pakistan's government, the Netherlands sent Jos de Sonneville to help. A tall, lean engineer with craggy features, water is de Sonneville's passion. He was 17 when he started sailing and had been a rower at university in Delft, where he studied civil engineering. The 1953 floods happened when he was six and although his family lived some distance from the hardest-hit parts of the country, he recalls his parents being stunned by the loss of life.

De Sonneville had spent much of his adult life overseas, advising countries in Asia and Africa on how to conserve and manage supplies of groundwater. In 1984, he witnessed first-hand the impact of catastrophic flooding in Mozambique, where he helped rescue people from treetops after a cyclone deluged the country. His time abroad shaped his understanding of disaster. Like Rousseau after the Lisbon earthquake of 1755, he didn't blame nature alone and believed that human intervention had made flooding worse.

In Pakistan, de Sonneville found that the rain and meltwater that used to sweep from mountain to sea now lay stagnant on the ground for months, in a delta where human settlement had left the river no space to expand.

Astonished onlookers videoed the moment when a torrent of rainwater swept down the Swat River and smashed through a luxury hotel in Kalam, a tourist resort in a steep, pine-forested valley high in the mountains of north-west Pakistan. Video of the New Honeymoon Hotel crumbling went viral after the floods. The hotel's glass windows shattered first, before boulders carried in the swirling river punched through its walls.

But it should never have been built so close to the water.[2] The hotel had already been destroyed once by flood, in 2010, only to be rebuilt in the same perilous spot. Alongside it, many more guesthouses were built at the water's edge. When the Swat flooded again, there were many more buildings in the river's path.

Three decades after man had begun to make peace with nature, Pakistan's floods underlined the disastrous consequences if humanity failed to treat water with respect. Through painful experience, the Dutch had learned to make room for their rivers, but, in 2022, that lesson would be brought home with deadly urgency in the cradle of one of the world's oldest civilisations.

From above, Pakistan appears to be a vast and arid plain, with a thick green line running down its centre, where the Indus River falls from the tallest mountains on Earth down to the Indian Ocean. The Indus is around 3,200 kilometres long, roughly half the length of the Nile, but carries twice as much water as Egypt's river. The search for water dominated the existence of early humans and the first cities were all nourished by great rivers. Modern humans are accustomed to thinking of floods as deadly and inconvenient, but the seasonal floods of the Indus sweep sediment and seeds down river, as well as creating shallow pools for fish to spawn. The river is life.

The city of Mohenjo-daro was built on the banks of the Indus around 2,600 BC, younger than the great cities of ancient Egypt and Mesopotamia (modern-day Iraq), but still one of the world's earliest civilisations. Its people were sustained by wheat and barley, and had streets laid out in a grid plan, as well as drains and sewers that kept their city clean. An elegant statue of a bearded man, his eyes half-closed in contemplation, was nicknamed 'the Priest-King' by archaeologists, though there is no evidence that he was either priest or king. Indeed, if this metropolis had an elite, they did not seem to require the palaces or temples of other ancient cultures.

The discovery during archaeological digs of unburied corpses in upper levels of the city, twisted together as if their deaths were sudden and violent, has given rise to the theory that the city was taken by conquest.

The British archaeologist Sir Mortimer Wheeler suspected invaders from the north had destroyed the civilisation that built Mohenjo-daro. He blamed a light-skinned Aryan race who worshipped Indra, a warrior-deity who brings thunder and storms, a cousin to those lightning-flingers Thor and Zeus. 'Indra stands accused' of the massacre at Mohenjo-daro, Sir Mortimer wrote in 1947.[3]

But archaeological digs had also found thick deposits of alluvial clay – clay, silt and sand deposited by a river – at various levels in the city ruins. In the 1960s, George Dales, an American archaeologist and former Marine, discovered that abandoned rooms and alleys in lower levels of the city had been filled up with rubble and dirt, in what he concluded was the last of 'several attempts to artificially raise the level of the city to keep above the height of the floodwaters'.[4] Bands of raiders may have taken advantage of chaos following the floods, Dales surmised, but they were not the cause of the city's decline. Indra was innocent.

The higher levels of the city were more shoddily built than its sophisticated lower levels; its final inhabitants used plain pottery rather than the elaborately painted vessels of their ancestors, creating a picture of squatters clinging on grimly in a world of ecological ruin. Geologists suggested that sea level rise was to blame, raising the level of the Indus upstream and triggering floods that eventually overwhelmed the city. The river is death.

In a land where rainfall, for most of the year, is sparse and unpredictable, canals that diverted water from the Indus enabled farming to flourish far from the river's banks. The Indus existed in people's imaginations as a living being, with village tradition passing on knowledge of the river's temperament and moods, with every bend host to a deity or saint.

Some of those stories live on into the present. Daanish Mustafa, an academic researching on a tributary of the Indus in the 1990s, was told that a stretch ran straight because an angry god, frustrated by the absence of his consort, had picked up the water and slammed it into the earth.[5] 'Every centimetre of the river has a story,' Mustafa said.

From the Middle Ages onwards, the land that is now Pakistan was governed by Muslim empires, with citadels on high ground (some with gates high enough for an elephant to pass through with an emperor on its back) and herdsmen tending goats and sheep, and rearing crops of millet and sorghum. Rain and snowmelt swelled the Indus and its tributaries during the *kharif* season, from the first rains in June to the end of the monsoon in September. Floodwater ran down natural channels that spread from the rivers, carving a winding path along the contours of the land, as well as man-made canals dug by farmers.

For the rest of the year, the *rabi* season, the Indus fell to a fraction of this volume. The canals filled with water and serviced fields for irrigation only in the peak months of the *kharif* season, when the river was fattest. The rhythm of agriculture was the rhythm of the river. Water is a living force that will silt up canals and break banks to carve new paths, so while local lords could guide the water flowing through their own territory, they needed a powerful state to manage the vast river.

When the British annexed the territory in 1843, they grasped the value of the existing canal network and extended the system. Their rationale was not just to increase the land's bounty, but to bind people to imperial rule. Sindh, the province where the Indus River flows out to the sea, was the western frontier of Britain's Indian empire. Here, nomads who had once moved flocks of sheep and goats to accommodate seasonal floods could, and in British minds *should*, be tied in place with grants of land to make them easier to oversee. 'Rude races first learn civilization by becoming possessed of property,' a British colonial official wrote in 1852.[6] Power over water was power over the people who depend on it.

There was another imperial consideration. The supply of water in the Indus Valley, limited to the final months of the *kharif* season, did not last long enough to get the best yields from a crop of cotton, the white thread that wove the entire global empire together, from the Atlantic slave trade to the mills of Manchester. Year-round irrigation could increase the harvest of a commercially valuable crop.

Richard Burton, a British intelligence officer with a gift for languages, explored Sindh in disguise as a Muslim merchant, creating a detailed impression of the province and its people which gave due prominence to the river. He called it 'the great fertilizer of the country, the medium of transit for merchandise and the main line of communication'.[7] Burton describes its floods in the *kharif* season as though it was an inland sea, with gyrating currents, rapids and shifting sandbanks. And he noted its nomadic nature, how the river shifted its bed in the course of successive seasons. The medieval capital of Thatta had once been washed by the Indus, bringing 'wealth and traffic', and was then neglected by the river, the city dwindling into insignificance as the wilful water brought prosperity to another settlement instead.

In 1855, another British officer, James George Fife, a lieutenant in the Royal Engineers, proposed a vast scheme to remodel the whole Sindh canal system. Fife was 30, but had already served with the army engineers for 11 years and, like Cornelius Vermuyden in the East Anglian fens, he had no shortage of self-confidence. His scheme involved cutting giant canals from the banks of the Indus, which would supply water even when the river was at its lowest. Fife argued that native people were not lazy, as some fellow British officials believed, but that the uncertainty of the river's flow made cultivation a lottery.

The scale of the young officer's ambition impressed his superiors, but the cost appears to have daunted them and he could not secure permission to carry out most of his works. Fife's ideas caught the attention of Florence Nightingale, the nurse and social reformer who had become famous for improving the hygiene of British military hospitals during the Crimean War. Nightingale never visited India, but she wrote to the officer about his work and publicly championed the cause of irrigation in India, believing it was one way to cure the famines that broke out periodically under British rule.[8]

Fife died in 1894, but his proposals were revived early in the 20th century when the British approved the building of the Sukkur barrage, a giant dam to control the Indus, accompanied by systems of canals on each bank of the river. This project, first outlined by Fife, would make Sindh, in the words of one official, 'one of the largest cotton-growing districts within the Empire'.[9] Completed in 1932, the barrage is a marvel of engineering, a wall of masonry spanning a 1.5-kilometre width of the Indus with 66 steel gates to control the flow of the river. The barrage supplies water to a 10,000-kilometre network of canals. This was an era of vast imperial water control projects – including a dam across the Nile at Aswan and a diversion of the waters of the Blue Nile to irrigate cotton fields in Sudan – but the reshaping of the Indus was the most extensive of all.

The swamps and mudflats of the Indus delta were thriving with wildlife, including the giant snakehead (a fish capable of growing to more than a metre long), the marsh crocodile and the Indus river dolphin. Damming and controlling the river reduced the flow of silt that brought fertility to its lower reaches and fragmented the habitat of animals. The Sukkur barrage was a concentration of power too, taking control of the water into the hands of government officials who could decide how much of the river to let through.

Just as in the Fens of eastern England, the domestication of nature represented a form of progress. Year-round irrigation vastly expanded the area that farmers could cultivate and meant they could grow crops in the *rabi* season as well as the *kharif* season. Growing commercial crops like cotton brought wealth to the empire, but it made farmers prosperous too. A farmer who could sell part of their crop for cash could use that money to improve their lives and buy food in leaner times. The largest swathes of newly irrigated land were granted to those who proved their loyalty to the empire; Khem Singh Bedi, a Sikh leader who helped the British put down an uprising, was awarded a holding of more than 3,000 hectares.

The nomadic river would be compelled to abandon its wandering ways, but early on, even among British observers, there were fears of the consequences. A river surrounded by embankments will carry sediment further downstream, gradually raising the riverbed and the level of the floodplain above the surrounding countryside. In 1849, a British officer warned that 'after a time there may be danger in retaining an immense flood above the level of the country'.[10] Engineering the river to fit human needs ran the risk of trading regular and modest floods for rarer but more catastrophic ones.

When two new states were carved out of the Indian subcontinent in 1947, the boundary line between India and Pakistan split the waters of the Indus, which rises in Tibet and flows through Kashmir before descending to the sea through Sindh. The most ancient rivalry in the world is the fight over water: the word 'rival' has its roots in the Latin word *rivalis*, meaning people who share access to the same stream.

But rivers themselves do not care for politics. They wind across borders as they seek the easiest way to the sea. In 1951, David Lilienthal, a US government official, published a plan for the countries to share management of the Indus. Lilienthal headed the Tennessee Valley Authority, which was created in 1933 to control floods and bring power and development to one of the poorest parts of the US, and he urged India and Pakistan to cooperate regardless of polit-ical differences.

After years of dispute, in 1960 the two nations agreed to the Indus Water Treaty, giving the Indus and two of its west-ern tributaries, the Chenab and the Jhelum, to Pakistan while India had use of three eastern rivers that flow into the Indus. But rather than the cooperation suggested by Lilienthal, this was closer to the bitter split of family assets in a divorce. And because India controls the higher reaches of the Indus, it can hold water back behind dams and reduce the flow of the river down to Pakistan. When the two states turn hostile, water can be a weapon.

The parting of the Indus required a vast new panoply of dams and canals to divert and control the water. To replace the water that had been 'lost' to India, two new dams were built by Pakistan in the upper reaches of the Indus, so that water could be penned up when it was abundant and released downstream when needed. One of these works, the Mangla Dam on the Jhelum River, is one of the world's biggest – more than 3 kilometres long and nearly 150 metres high.

Its creation in 1967 submerged the old city of Mirpur and dozens of surrounding villages, displacing thousands of people. The dam was a symbol of progress, generating vast quantities of electricity as well as controlling the flow of water for irrigation. In the eyes of the government, the cost of displacing people paled beside the benefit to the entire nation. (Many of those who were displaced emigrated to Rochdale, Bradford and other northern English towns and cities, following an earlier wave of migrants who were drawn by the prospect of well-paid work in British factories. Their descendants make up a substantial proportion of Britain's Asian community.) Alongside these dams and irrigation canals, channels were constructed to provide flood relief, with the aim of draining the waters of the Indus away from fields and villages and into the Arabian Sea.

A river that is confined behind dams and levees, and whose flow is directed along canals, becomes less visible. That is literally true – and true as an idea, too. Myths of the river may be handed down, but practical understanding of how to manage a river's changing moods becomes less valuable than the expertise of engineers. Before the river was carved up for agriculture, the floods of the Indus would be wide but shallow. Now, an empire of dams and canals blocked the advance and retreat of floodwater. The purpose of the system was to extract the maximum harvest – and profit – from the land. Disaster came as a by-product.

In August 1973, heavy rains in the mountains of Kashmir combined with snowmelt to turn the Indus into an angry torrent. The swollen river surged south through Sindh, crumbling the earthen levees meant to constrain its path and bringing devastation, flooding towns and hundreds of villages and ruining crops. It was the worst flood in the young nation's history and the prime minister Zulfikar Ali Bhutto, an Oxford-educated barrister who dominated his country's politics for decades, described it as a 'national calamity'.[11]

But it was one calamity in a long list of weather-related disasters around the world in the 1970s, to which analysts at the CIA were paying attention.[12] A working paper completed in August 1974 noted that one country after another was experiencing the impact of a changing climate. The headlines, the report's authors wrote told 'a story still not fully understood or one we don't want to face'. Climate was becoming a critical factor, the analysts stated, because it threatened the supply of food, on which the stability of nations depends.

Understanding that the climate was altering, and that humans were responsible, had reached the highest levels. A group of scientists had warned the US president Lyndon Johnson in November 1965 that humanity was conducting a vast experiment, burning in a few generations the fossil fuels that had accumulated in the earth over hundreds of millions of years.

Among the experts was Charles Keeling, the researcher who had taken measurements of atmospheric CO_2 on the slopes of Mauna Loa. In the early 20th century, Svante Arrhenius had theorised that the climate could be transformed as a result of the gases spewed out by industry, but Keeling had gathered the data which measured just how much CO_2 was accumulating invisibly in the atmosphere.

The experts warned the president that warming of the atmosphere might result in a melting of the Antarctic ice cap, rapid by geological timescales but slow from a human perspective, which would in turn raise sea levels. By the late 1970s, executives at the biggest oil company, Exxon (now ExxonMobil), knew the world was changing. James Black, a scientist at the company, made a presentation to Exxon executives in July 1977, in which he summarised the science.

Black warned the oil company executives that a warmer world was on its way, with more rainfall, in which both deserts and fertile areas of the Earth might shift to higher latitudes. When Keeling had begun taking his measurements, carbon dioxide was at 313 parts per million in the atmosphere. By 1977, it had reached 330 parts per million. There was still time to act – a window of five to ten years, Black said – before the need for hard decisions about the energy humanity consumed would become critical. In public, however, the company emphasised what it described as 'fundamental gaps' in knowledge about the climate.[13]

By the late 1980s, governments were increasingly focused on the risks of climate change, grappling with how they would reduce greenhouse gas emissions while adapting to a rise in sea levels and changing rainfall patterns. The Dutch convened the first international climate talks in the resort town of Noordwijk, on the North Sea coast, in November 1989, where countries agreed on the need to stabilise emissions from industrialised nations.

But the US, Japan and the Soviet Union joined forces to veto the crucial step: setting targets. It set the pattern for decades of climate negotiations, which were consistently frustrated by wealthier countries, until, in 2015, 195 countries reached an agreement in Paris to lower emissions and keep the global rise in temperatures 'well below 2° C'; that is, below a 2° C increase on the average temperature before humanity began pumping out carbon dioxide in vast quantities.

The Paris Agreement was a good deal for the world's richest countries. Previous efforts to deal with climate change had focused on them, but this treaty called on every signatory country to take some form of action. It was far from ideal for the world's poorest. In particular, there was no legally binding commitment on rich nations to support poorer ones with the costs of adapting to a more unruly planet.

But the leaders of small islands – from those built on coral atolls, such as the Maldives, to volcanic islands like Montserrat in the Caribbean – had fought for an agreement that limited the rise in temperature to below 1.5° C. Above that threshold, their countries faced severe risks from sea level rise and coastal flooding. That made the breakthrough in Paris a narrow victory for the whole species.

It was a victory on paper at least. The next ten years saw a weakening of political will and vast quantities of carbon being pumped out into the atmosphere. With the exception of 2020, when factories shut down and planes were grounded by the pandemic, carbon emissions surged to record highs. The year 2024 was the hottest on record. There was some good news: the cost of solar and wind power fell sharply, displacing some fossil fuel use. But in developing countries like India, the growth of renewable energy did not keep pace with a rising population that had growing demand for energy.

When Sherry Rehman was appointed Pakistan's climate change minister in April 2022, the air conditioning in government offices was humming at full pelt while, outside in the streets, the heat was stifling. Pakistan is a hot country, but this smothering intensity was something new. Schools were closed, people kept to the shade when they went outside and bathed as often as they could. The temperature soared above 40° C in parts of the country. Rehman told an interviewer: 'When you have an apocalypse in front of you … Have you not watched Hollywood movies? You have to face it head on.'[14]

Rehman had experience of facing challenges head on. She was one of a handful of women in a male-dominated political system, a former newspaper editor inspired to enter politics by Benazir Bhutto, Pakistan's first female prime minister. She joined Bhutto's Pakistan People's Party, the centre-left party founded by Bhutto's father, and the two women shared a style: Pakistan's traditional *salwar kameez*, flowing tunics and trousers with a matching *dupatta* shawl wrapped around their shoulders and designer eyeglasses. They shared toughness too. When Bhutto's convoy was attacked by a suicide bomber in 2007, Rehman was in the same car. She still has the scars.

Weeks later, when Bhutto was attacked again, by a gunman and suicide bomber, it was in Rehman's car that she was rushed to hospital. That time, Bhutto did not survive the attack. In a country wrestling with increasingly assertive and violent religious extremists, Rehman herself faced death threats for her opposition to Pakistan's blasphemy laws. If that rattled her, she never let it show.

Dozens died from the heat in the spring of 2022, while wheat crops were stunted by the unseasonably high temperatures. But the country was in political turmoil too. Imran Khan, a former cricketer who had used his star power to fuel a political campaign, had just been ousted as prime minister, losing a vote of no confidence. Khan had promised to clean up corruption, but struggled with the challenges of a government that was deeply in debt and an economy in the grip of soaring inflation. And he appeared to have lost the trust of the country's military. He was replaced by the leader of Rehman's party. While the country was gripped by Khan's downfall, the second act of Pakistan's climate disaster began to unfold.

In the ice and rock of Pakistan's high north, the unusual spring temperatures had accelerated the melting of glaciers, with water pooling up in glacial lakes, bodies of near-freezing water that are given a brilliant turquoise colour by the fine silt suspended in them. These lakes are penned in by the ice and debris that pile up at the toe of a glacier, until the turquoise water rises high enough to burst over this natural dam and the flood cascades down onto the valley below. The normal seasonal rhythm is for a glacier to gather snow in winter and release it through the spring and summer, a natural store of fresh water that helps nourish agriculture in the plains. But a sudden surge in temperature upsets that balance.

On 7 May, a lake at the foot of the Shishper glacier in northern Pakistan overflowed. A torrent of water, now brown with the mud and debris it carried, surged down the hillside and slammed into the giant concrete supports of the Hassanabad Bridge, which spans the River Hunza on the mountain highway between Pakistan and China. Part of the bridge collapsed into the river and was rapidly swept away. Pakistan has more than 7,000 glaciers, more than any other country outside the Arctic Circle, and they had become swords of ice, poised above the heads of its people. Rehman shared an image of the bridge collapsing and linked the disaster to the species-wide failure to act on climate change. 'We need global leaders to reduce emissions, walk the talk,' she posted.

Between July and August, Pakistan was hit by three unusually strong rainstorms, a record-breaking deluge. A total of 45 centimetres of rain fell in the space of three months, four times the average for the season. Floods do not just take lives and wreck homes; they also erase memories of previous disasters. The rain that pummelled the country in 2022 was twice the volume that fell in 2010, the last time that a flood devastated Pakistan. Late in August, with hundreds dead and vast tracts of the country underwater while the rain continued to pelt down, the government declared a national emergency. In interviews, Rehman called it a 'monster monsoon'.

By September, southern Pakistan was unrecognisable. Villages built of low-rise yellow-brick houses were marooned, or half-submerged; the fields that had once been green with cotton shrubs were now underwater.

The floods suspended village time. Until they receded, crops could not be planted and goods could not be trucked in and out. Mosquitoes rather than crops flourished in the stinking stagnant water. It's estimated that the inundation of 2022 caused economic losses of $15 billion, while the cost of reconstruction is reckoned at more than $16 billion.

Faiz Ali, a young villager, described swimming for 20 minutes along water that pooled where a road had been to fetch groceries from the nearest town, Johi. 'I'm still afraid every time I go,' he told reporters. When the flood hit Johi, the townspeople acted quickly to protect themselves, filling bags with stones and sand to shore up the embankment protecting the town. Their quick thinking meant it survived the worst of the flooding, but they felt abandoned by their government.

Floodwater gathering on land to the west of the Indus would naturally flow east into the river, but a gargantuan drainage channel carried it southwards instead, into Lake Manchar, an expanse of freshwater that attracts flocks of pelicans and spoonbills. As the lake swelled with floodwater,

Pakistani officials feared it would deluge nearby cities. In early September, authorities in the district deliberately breached the banks of the lake, diverting water away from cities but allowing it to submerge nearby villages. People were caught unaware as water rushed onto their land, destroying crops and leaving cattle up to their necks in water.

By 2022, the world was 1.2° C hotter on average than it had been in pre-industrial times. The burning of fossil fuels had led humanity close to the 1.5° C threshold and the fingerprints of man-made climate change were all over this disaster. The exceptional heatwave, followed by the overwhelming monsoon, were both made more likely by climate change.

De Sonneville put together a team of Dutch experts who visited Pakistan to advise on how future flood disasters could be prevented. The team found that railway lines blocked the flow of the water. Rail lines had been built on earthen embankments to provide a stable and elevated surface, but these raised earth ridges did not have sufficient drainage to let water run off and instead trapped it in farmers' fields. The only way the floodwater could now escape was the slow process of evaporation.

There was a further problem. Identifying and understanding the precise nature of flood risk in each location required detailed mapping of the terrain in order to track the gradients across which water would move and identify the low points where it would pool. But, in Pakistan, this data was held by the military. One of the Dutch experts said: 'The military would say "Thank you, we have that data – we're not going to share it with you."'[15]

169

Officials in Pakistan's four provinces did not collaborate to manage flood risk. Instead, there was competition and friction between them, with lower provinces accusing higher ones of holding water back behind dams and barrages when there was a drought – and sending water downstream too quickly when there was an excess. This was not a problem unique to Pakistan; many countries treated rivers as if they began and ended at the borders of provinces. If you want to manage water, the Dutch team told Pakistani officials, you cannot do it without dialogue.

Pakistan's leaders had been slow to take climate change seriously as they grappled with political turmoil and a worsening economy. Russia's invasion of Ukraine that February had driven oil and food prices higher around the world, leaving Pakistan's economy in crisis. That year, Rehman's climate ministry had been granted a modest budget – just $43 million out of a national government budget of $47 billion.[16] The sum was clearly dwarfed by the vast environmental changes now taking place. But Rehman now had a rare opportunity to make her case to the world.

The year before, Pakistan had been elected to lead the Group of 77, the main negotiating bloc for developing countries at UN climate talks. The world's leaders were preoccupied by the war on Europe's eastern flank, while, at home, the governments of rich democracies faced rising support for right-wing populist movements who scorned the need to tackle climate change, focusing instead on securing their own borders against migration. When it came to cutting the carbon dioxide being pumped out from burning fossil fuels, the international will to act was dribbling away. In fact, global emissions from energy and industry had risen to a record high in 2021. Things were looking bleak.

Pakistan's tragedy became the emblem of the world's disarray on climate change. Over footage of villages now mired in green floodwater, Rehman declared that a third of the country was underwater. Although satellite images indicated the area of flooding was closer to a tenth of Pakistan, there was no doubting the severity of the disaster.

Rehman found an ally in António Guterres, who was now in his final term as UN secretary-general and determined to push for action on a dangerously heating planet. The world's chief diplomat visited Pakistan in September 2022 and declared: 'Humanity has been waging war on nature and nature strikes back.'[17]

As the causes of climate change became clearer, developing countries facing the worst impacts of rising seas and violent hurricanes have, for three decades, pushed richer nations (who have historically burned most oil and gas) to provide compensation. Wealthy countries have resisted that demand, fearing the scale of their potential liability. However, when more than 200 nations met at the Egyptian seaside resort of Sharm El-Sheikh for the UN's climate talks that November, there was an unexpected breakthrough.

Confronted by images of Pakistan's disaster, leaders of rich nations agreed to create a climate damage fund. It was, Rehman said at Sharm El-Sheikh, 'a down-payment and investment in climate justice'. But the talks failed to agree on stronger measures to curb the soaring growth in greenhouse gas emissions. The price of crude oil had soared after the outbreak of war and governments everywhere were anxious about the cost of energy. There was no agreement on phasing out oil and gas. Instead, the world was now on track for between 2.1 and 2.9° C of warming above pre-industrial temperatures.

And there was still no answer to the recurring disaster of Pakistan's floods, which were caused not just by climate change but by the country's failure to live with the moods of the Indus. In the Netherlands, there was space to yield some land back to the rivers, but Pakistan's population had grown from 33 million people at independence to more than 240 million people by 2022, making it the fifth most populous country in the world after India, China, the US and Indonesia. More than half of that population lived in the countryside.

De Sonneville recommended modelling what happened when the country flooded and using this model as a basis for negotiation between provinces, so they could draw up a collective plan to handle future crises. But before these recommendations could be adopted, there was an election. In an astonishing twist of Pakistan's political drama, candidates aligned with Imran Khan won the most seats in that vote in February 2024, but not enough to form a government. Instead, his rivals pieced together enough votes to form a shaky coalition.

The outcome was a setback for the military, which has long been seen as the controlling force in Pakistan's politics. But it was also just the latest indication of how overlapping political and economic crises were paralysing the world's efforts to deal with climate change. Like many of the world's poorer countries, it is dependent on hydrocarbons. As European countries sought to reduce their dependence on Russia, they scrambled to buy liquefied natural gas from the US and Qatar, pushing up the price. Priced out, Pakistan was forced to generate electricity by burning coal, the most polluting of all fossil fuels.

Barring a miraculous transformation in the world's politics, it appears unlikely that humanity will bring emissions down fast enough in the near future to prevent the planet heating significantly. Our future challenge will be with managing water, rather than CO_2.

The giant irrigation system surrounding the Indus was built in an era when it was thought water would always be plentiful. Beginning under the British, farmland was supplied with wastefully abundant quantities of water. In good times, this was profitable, but the consequence of ruthlessly exploiting the environment was creating a system vulnerable to flood disasters. Giving land back to create a more naturally flowing river would always be difficult when it ran counter to the interests of wealthy landowners.

Climate change has created a world in which water is both scarcer and more plentiful, with parched and searing dry seasons turning to overwhelming monsoons. Managing water under these conditions is not an impossible problem to solve, but it may require shifts to less thirsty crops or more efficient forms of irrigation.

The river has always brought both life and death. Climate change just alters the scale and frequency of its floods.

Rising to this challenge will require a shift in political thinking. But the increasing quantity of CO_2 in the atmosphere is not just transforming the environment. It is beginning to alter politics too.

CHAPTER 8

STORM AUTOCRATS

In the middle of the night in September 2023, the storm pummelled two crumbling dams above the Libyan city of Derna, saturating them both with water. The dams, built from clay and rock in the 1970s under dictator Muammar Gaddafi, had been poorly maintained and were riddled with cracks.

The higher of the two was the first to shatter, water cascading down onto the lower dam, which collapsed in turn and sent a mud-brown torrent of water, rock and rubble hurtling down from the hillsides. The flood slammed into Derna with a force that swept away entire neighbourhoods. At least 4,000 people were killed according to the official death toll, though thousands were unaccounted for, their bodies swept out to sea and never recovered.

Dams usually symbolise man's subjugation of nature, holding back immense volumes of water to serve the needs of farming and industry. For this reason, they're frequently an emblem of unaccountable power, the favoured mega-project of dictators and one-party states. But the shattering of the dams above Derna raised a new and troubling question:

whether disasters could also create the ideal conditions for an autocratic ruler to consolidate his grip on power.

As volunteers and aid agencies trawled through the rubble of the city for survivors and distributed blankets, food and medicine, they worked surrounded by militiamen loyal to the Libyan warlord Khalifa Haftar. His youngest son Saddam took official charge of relief efforts, though he had little experience of disaster relief. Libyans knew him better as the head of an armed group that had tortured and murdered critics of the military.[1]

Doubts about Saddam were strengthened when a British television journalist asked him if the floods could have been prevented.[2] From the window of his pick-up truck, he dismissed any criticism with a scowl and the curt reply: 'Everything's fine.'

Derna, a seaside town in eastern Libya, was known as the 'city of poets', famous for its cafés and bookshops, as well as for lush gardens that produce delectable figs and oranges. Older generations of Libyans said that a piece of heaven had fallen from the sky. One of the dams that crumbled, Bu Mansur, is 13 kilometres south of the city and perched high over a ravine in the Green Mountains above Derna. The ravine is deep and dry, a few green shrubs dotted on its steep sun-baked sides, until the rainy season late in winter when flash floods sweep down the valley and spill onto the city.

The high dam and its lower companion were built by a Yugoslav company in the 1970s to protect the city from floods and store water for irrigation at a time when Gaddafi preached that the country should be self-sufficient in agriculture. A storm in 1986 left them damaged, but still standing. Even under Gaddafi's regime, Libyan authorities were aware of the declining state of the ageing dams, hiring a Swiss consultancy which found potential structural flaws. A year before the dams were breached, a Libyan civil engineer, Abdelwanees Ashoor, warned that officials needed to take immediate measures, writing in a research paper that 'in the event of a huge flood, the result will be disastrous'.[3]

The flood, when it came, was mercilessly swift. At sunset, the bottom of the ravine was still dry. The upper dam's security guard, who lives in an isolated house on a cliffside above the ravine, contacted his supervisor by phone at midnight to warn him of the rising water. After that, he could not get through again. Between midnight and 2 a.m., the guard observed the water rising at speed. Then it began pouring over the crest of the dam. As the rocks split apart under the weight of the floodwater, the friction between them created a smell of burning.[4] The upper dam gave way around 2.40 a.m.

The scale of the rainfall was far beyond any previously recorded storm in Libya. The volume of water in the ravine was later estimated to be seven times greater than the dams' capacity to withstand.[5] Dams that had been better maintained might have endured the pent-up water for longer, allowing more time for evacuation, but even if the barriers of clay and rock had been structurally sound, they would ultimately have crumbled in the face of the torrent. What happened was worse than if there had been no dams at all. The sudden collapse in the early hours of the morning amplified the disaster, unleashing an overwhelming surge of water onto the city below.

The storm was a Mediterranean cyclone, Storm Daniel, which had brought torrential rain and flooding to Bulgaria, Greece and Turkey, claiming more than a dozen lives in the three countries. As it crossed the Mediterranean to the north coast of Africa, it gained strength from the sea's warm waters.

Libya's national meteorological centre had given notice of extreme weather 72 hours before the storm hit, while Derna's mayor had warned residents to evacuate parts of the city – even though he believed the storm threatened coastal parts of the city rather than areas further inland. Few residents left, and military authorities then ordered a curfew.[6]

At first light, the devastation was breathtaking. A 6-metre high wall of water had shattered many of the city's concrete buildings, sweeping some away completely while leaving others as heaps of rubble. Crumpled and overturned cars littered the streets along with the traces of human habitation: scattered clothes, shoes and children's toys.

The disaster took place in September, just a year after the monsoon floods that inundated Pakistan. It would have been hard to bear in a country that was well-run, but Libya was far from that. After a revolt during the Arab Spring of 2011, which brought Gaddafi's dictatorship to a violent end, the country was split between rival armed factions. Khalifa Haftar, a rogue general in the Libyan military who was recruited by the CIA in the 1980s to be a thorn in Gaddafi's side, carved out a fiefdom in the east of the country, including Derna.

Haftar had attempted to seize the capital, Tripoli, with the support of Russian mercenaries in 2019, but the general's forces were beaten back after Turkey intervened, supplying armed drones that rained missiles on his bases and supply lines.

The flood brought an outpouring of solidarity. Trucks graffitied with the names of western Libyan towns arrived in Derna with aid. Turkey, despite its military intervention to support the other side in the war, was one of the countries that sent aid and a team of rescuers. There was a glimpse of reconciliation, but in the ruined city, grief was souring into anger. Failure to maintain the dam, combined with the lack of evacuation, had resulted in catastrophic loss of life.

A week after the disaster, with many neighbourhoods flattened and a trail of mud and debris marking the flood's devastating path through the city, hundreds gathered outside the Sahaba mosque, a city landmark. The protesters, some of whom climbed onto the flat roof around the mosque's golden dome, called for the resignation of Derna officials. Reporters gathered to film the protest, including Libyan TV crews that broadcast their demands across the country. Other protesters marched to the mayor's house and set it on fire.

Soon after, internet and mobile networks were switched off across the city. Although the mobile signal had survived the flood, the official explanation was that fibre-optic cables had since been severed, disabling the networks. The square around the Sahaba mosque was barricaded by soldiers. Journalists were asked to leave, told that their presence was hampering the efforts of rescue workers. These explanations might have had a cloak of plausibility, but the timing suggested a desire to squash awkward questions.

A few days before the protest at the mosque, internal security agents detained a man who had given TV interviews in which he criticised the authorities for their lack of preparedness before the floods. The arrests escalated after the demonstration. According to evidence gathered by Amnesty International, one of the protest organisers was snatched from his home by armed men who clubbed him with the butts of their rifles before hustling him into their vehicle and driving away.[7] Libyan authorities opened an investigation into the disaster and Derna's mayor was arrested and taken into custody.

When the flood hit, Khalifa Haftar was 79 and considering who would succeed him. On the day of the disaster, his eldest son Elseddik announced his ambition to run for president. By the end of the year, another of the general's sons, Belgacem Haftar, had been appointed head of Derna's reconstruction fund. The disaster set a pattern: dissent would be quashed, lower-ranking officials would bear the blame and a strongman would emerge stronger, prepared to make his fiefdom a family affair.

'It has left them in a better place,' said Anas el-Gomati of the Libyan thinktank the Sadeq Institute.[8] 'They have consolidated power in eastern Libya. They have become increasingly wealthy.'

Libya is a divided nation, existing in a precarious balance between competing governments in east and west, but it may offer a glimpse of a dystopian future. Already, less than half the world's population lives in a democracy of some kind, a share that has fallen over the past decade.[9] A few years ago, a group of economists began to wonder whether there was a link between the shock of a natural disaster and the rise of authoritarianism.

Rafael Trujillo, the general who ruled the Dominican Republic for three decades, is one of the earliest examples of a leader who seized absolute power in the wake of a storm. When Hurricane San Zenón struck the Dominican Republic in 1930, ripping roofs off houses in the capital city and killing thousands, Trujillo placed the country under martial law and took charge of handing out aid. He rebuilt the storm-damaged capital, renaming it after himself: Ciudad Trujillo. By the time martial law was lifted in 1934, Trujillo had murdered hundreds of political opponents and cemented his control.

One possibility is that men like Haftar and Trujillo represent isolated examples of strongmen who rode the wings of a storm, simply opportunists who saw their chance to seize greater power. But the more disturbing thought is that there might be a pattern here. Habib Rahman, an economist at Durham University, first became interested in the link between disaster and politics in Bangladesh, his native country. Before moving to academia, he worked with governments – including his own – on disaster preparedness.

Together with colleagues at Durham and at Deakin University in Australia, Rahman focused on island states, as storms affect a larger proportion of their territory compared to bigger countries. The researchers' starting point was that storms were a shock that could create a political tipping point. As in Libya, the risk facing rulers in storm-hit countries was that citizens would protest if their concerns were not listened to. Governments could pacify their people by taking charge of aid provision, but the temptation was to tighten political control at the same time.

Rahman and his colleagues compared islands that are frequently exposed to storms, like the Dominican Republic and Fiji, with islands like Singapore and Iceland that are sheltered from extreme weather. Ranging across seven decades of data, they found that in the year after a storm, political repression intensified and constraints on leaders grew weaker.

Trujillo had only been the first of his kind, the researchers suggested. The rise of authoritarian regimes in countries ravaged by frequent floods marked the coming of a new political model: the storm autocracy.[10] When disasters strike, well-intentioned foreign governments rush to give assistance to survivors, but the researchers found that democracy in storm-hit countries deteriorated after disaster aid was given. Emergency funds were helping to strengthen the grip of tyrants and, as the planet warmed up and countries experienced more frequent disasters, storm autocrats would become more powerful.

It's not the first time a link has been made between autocracy and the environment. In the late 1950s, the German historian Karl Wittfogel put forward a theory that, in ancient civilisations, building and operating irrigation systems and flood defences fostered the rise of a despotic and bureaucratic state.[11] Wittfogel acknowledged that a single farmer or community could engage in small-scale canal building, but proposed that in semi-arid landscapes like ancient Egypt, where farmers could not regularly rely on the rain, government control of water produced what he described as 'hydraulic civilisations'.

China's flood myth tells the story of the emperor Yu, who is said to have dredged river beds and built irrigation channels to harness river floods and make them serve the needs of farming in a labour of love that took many years. Yu, a pioneer of flood control, became the founder of a dynasty. Too little or too much water did not necessarily lead to state control, Wittfogel acknowledged, but above the level of subsistence farmers and below the level of a modern industrial society, his theory was that the need to provide disciplined labour to channel large quantities of water and protect crops from floods tipped societies towards despotism.

The timing of Wittfogel's idea was significant. The historian had been a member of the Communist Party in Germany, but suffered the Marxist version of a crisis of faith after the Soviet Union agreed a non-aggression pact with Nazi Germany. His book *Oriental Despotism* was first published in 1957; his criticism is aimed at communist China and Russia as much as it claims to analyse the ancient world.

Wittfogel's theory risks slipping into over-simplification: the argument that absolute monarchies had taken hold in the Middle East or in China primarily because of their environment implied that it had to be that way and human choices made little difference, that geography is destiny.

There are many exceptions to the theory. Ancient societies in places like Papua New Guinea developed sophisticated irrigation systems without any sign of a bureaucratic state.

Imperial powers like the Romans are often associated with feats of water transportation, but they frequently adopted water systems that had been developed locally, by earlier cultures; it wasn't the aqueducts that built their empire.

In more recent times, the example of the Netherlands suggests that water management and flood prevention can be a spur to cooperation and the birth of democracy. In contrast, while the British tried to strengthen imperial power with the irrigation works they built in the Indus basin, they were forced to give up their Indian empire a few decades later. There's a more subtle way to interpret this: environments shape societies, and people shape their environments in turn, and the effects of this interaction reverberate through history, shaping the choices made by future generations.

The British left India, but the structures they created around the Indus left a legacy. When a river is dammed and channelled, and roads and farmland spread in its basin, as happened in Pakistan, it becomes difficult to give that land back to nature and let the river breathe again. Once marshland is drained, as it was in eastern England, it needs to be managed. The land has subsided and now lies far below sea level, so will not revert naturally to marsh. If East Anglia is not continually pumped dry, it is more likely to become a landscape of permanent standing floodwater.

The reshaping of nature creates winners and losers. People who once lived in a place often end up dispossessed, but others profit from speculation on draining marshes. Elites have an interest in maintaining the system as it is, even if that makes the countries they live in more vulnerable to floods.

After all, it's not usually the rich and powerful who suffer the worst consequences.

Political choices are guided by philosophy too. Humanity's relationship with nature has always been one of give-and-take. From earliest times, there's been a philosophical understanding that the planet and its bounty were not just resources to harvest, but that they needed protection. There were months in which wild animals were not hunted because it was their breeding season, while fields were left fallow to allow soil to recover. But in modern times, societies have tended to take a more extractive view.

The exploitative impulse reached an extreme under the dictatorships of the 20th century. In China, opponents of dam building have been dismissed from government positions and worse. Mao declared that 'man must conquer nature' and under him both people and the natural world were subject to the arbitrary and unchecked exercise of government power. During Mao's rule, Huang Wanli, a hydrologist who criticised the building of a dam to provide flood control on the Yellow River, had his career cut short and was ordered to carry out forced labour on a construction site. The hydrologist was vindicated, as the dam became silted up by the river's sediment, rendering it useless.

Environmental destruction can be an act of repression in itself. After the first Gulf War ended in 1991, the Arabs of Iraq's southern marshland joined an uprising against the dictatorship of Saddam Hussein. For centuries they had navigated the green, muddy waters of their marshes in slender canoes, herding water buffalo and building elaborate reed houses known as *mudhif* that rested on islands of compacted mud. The marshes had always been a haven, a refuge for runaway slaves and deserting soldiers.

After the rebellion was defeated, Saddam ordered the draining of the marshes, officially for irrigation but, in practice, an ecological disaster that resulted in the land becoming barren. Many of the Marsh Arabs were arrested or executed. Others became refugees abroad or were forced to seek new homes within Iraq. A population that once numbered around a quarter of a million, along with a way of life so old that it was depicted on ancient clay tablets, was erased within a few years.

Shaping the environment to fit human needs generated a backlash that found creative and political expression from the 1960s onwards, from Rachel Carson's books to Neil Young singing of 'Mother Nature on the run'. In democracies, environmental activists were able to exert pressure on governments, as in the Netherlands, where van Veen's original design for the Delta Works was adapted to balance the threat of flooding with the potential harm to the Eastern Scheldt tidal estuary.

In the US, the green movement enjoyed extraordinary success in the 1970s, with the passage of major legislation protecting nature, including the Endangered Species Act, which safeguarded not only plants and animals but their habitats. The conservation law was passed with bipartisan support and signed into law in 1973 by a Republican president: Richard Nixon.

The new law's first test came when a lawsuit filed on behalf of an endangered fish, the 7-centimetre-long snail darter, delayed the construction of a dam in Tennessee. The fish won all the way up to the Supreme Court, but then lost when Congress passed legislation exempting the dam from the conservation law. (The story has a twist: the fish were collected and moved to other rivers in the south-east US, where they thrived. The species bounced back and is no longer endangered.)

Industry chafed under environmental regulation, warning that it was increasing costs and making it difficult to hire, though costs were often lower than feared and new jobs were sometimes created as companies hired staff or employed contractors to clean up pollution.

As climate change became increasingly prominent in the 1980s, lobby groups with ties to the oil and gas industry began to push back against government action on emissions. In 2001, under George W. Bush, the US withdrew from the Kyoto Protocol, an international treaty in which rich countries agreed to control their output of greenhouse gases.

The language and actions of US Republicans diverged from right-wing parties in Europe, including the UK, where Margaret Thatcher was one of the first to warn about risks of climate change and where Conservative governments committed to reaching net-zero emissions (although technology can sometimes be a more powerful force than politics; emissions have dropped in both the US and UK, chiefly because coal-fired power plants have been replaced by gas, wind and solar).

The real shift came when radical right parties began to achieve electoral breakthroughs in the first two decades of this century. Germany's Alternative für Deutschland attacked 'carbon hysteria which is structurally destroying our society, culture and way of life'.[12] Britain's Reform wants to block the expansion of renewable energy and focus on investment in North Sea oil and gas.

For the new breed of right-wing parties and their voters, the emphasis on attacking green policies is more cultural than economic. It forms part of a rejection of the 'woke' values of the university-educated liberals they disdain. Survey data from six countries – Brazil, South Korea, Japan, France, the UK and Canada – shows the highest correlation with climate-sceptic views is not anxiety about the cost of living or any other economic factor, but the belief that 'feminism has gone too far'.[13]

Social media, in particular Elon Musk's platform X, has helped propagate climate-sceptic views. Climate-sceptic influencers are increasingly pushing claims that policies aimed at reducing emissions, from eating less meat to flying less, are a form of elite control: 'climate communism'.

Increasingly, the centre-right has been dragged into the populists' orbit. In the UK, successive Conservative leaders made environmentalism part of their messaging, from David Cameron posing with huskies on a visit to a Norwegian glacier to Boris Johnson hosting UN climate change talks in Glasgow. The appeal may have been a tactical one, a way of connecting with voters who might otherwise be drawn to more left-wing parties. Sometimes, it was more personal;

Johnson's wife Carrie worked for a wildlife charity and is a passionate supporter of animal welfare. Under challenge from the populist right, the Tory party has now abandoned the goal that the UK should reach net zero by the middle of the century.

The shift in tone by the party's leadership has been matched by a shift in its voters. Until recently, Tory voters were in favour of net-zero policies, while Reform voters were opposed. Polling indicates the Conservative voter base is now split 50/50 on the issue. Steve Akehurst of the think tank Persuasion UK said: 'A couple of years ago, there were prominent Conservative voices who sent signals to right-wing voters that this is our thing. Now that signalling doesn't exist.'

The impetus behind this shift could have psychological roots. Researchers have found that a worsening climate has the potential to widen divisions between social groups. The study of climate's effects on the mind is a new field, but it builds on existing research which shows that, in a time of crisis, people are impelled to seek the shelter of their tribe. The sense of helplessness people feel in the face of planetary changes makes them want to reassert control, both in the safety of their in-group and the strength of a powerful leader.

Immo Fritsche, a professor of psychology at Leipzig University, conducted experiments with a group of white British people who were exposed to threatening information about climate change.[14] This included a map of Britain showing areas of the country threatened by sea level rise by the middle of this century. A control group was given bland facts about the country, such as the fact that the Severn is the longest river in the UK.

Fritsche and the other researchers then gave the participants a questionnaire about their mood over the previous week, with the idea of delaying their response. The idea is that a threat has its most powerful effect when it sinks to the unconscious level of the mind. Fritsche found that when people are exposed to threatening information about climate change, and also lack confidence in their own government's ability to deal with the climate threat, their attitudes to minorities become more negative compared to the control group.

This negative attitude was most conspicuous when it came to Muslims and Pakistanis, two groups that have faced a hostile press in Britain in recent decades. This might stem from seeing minorities as a threat to the in-group's ability to act together to achieve shared goals, Fritsche suggests. Setting a border around their community created a sanctuary where people could reassert control.

Defending against floods requires a degree of consensus. If the age of floods proves to be an age of autocracy, it will have troubling consequences for how well societies maintain flood defences and prepare for future disasters.

Maintaining a flood barrier requires cooperation across generations, preserving a high standard of maintenance, finance and technical knowledge. Referring to the barrier that protects Rotterdam, Marc Walraven, an adviser to the Dutch government, said: 'Every country can design and build this kind of storm surge barrier. The challenge comes after that.'[15]

Unlike a tunnel or a bridge, a flood defence can go years without proving its worth, but a single lapse can be disastrous. 'If there's a problem on a highway, you can easily lower the speed limit,' Walraven said. 'With a storm surge barrier, you can't say that today it is working at 50 per cent.'

The era of dam construction in the rich world has peaked, but vast dam projects are underway across Asia and Africa to provide both hydropower and flood control. As the size of storms increases with climate change, both new and existing dams will require more regular oversight to ensure they remain functional. But civil wars and weak governments bring the risk of further disasters like Derna.

The tragedy in Derna might have been a catalyst that brought Libya's two rival governments together to address the challenge of the ruined dams above the city. Instead, while massive reconstruction projects got underway under General Haftar's son Belgacem, there was no sign of reconciliation, nor a sign that the city was being rebuilt with an understanding of its vulnerability.

Libya has suffered from years of war, but it has oil wealth, which meant there were plenty of funds with which to rebuild. The city buzzed with construction machinery and building workers after the flood, with residents astonished by the speed at which new apartment blocks, schools and clinics went up. New bridges linking the east and west sides of the city were built to replace those swept away in the collapse of the dams.

The rapid restoration of the city left little time to discuss its future.

As both the Dutch and British understood after the 1953 floods, it's critical to link the risk of floods to the scale of death and destruction if a flood defence is breached. Instead of rebuilding houses in the path of a potential flood, urban land could be converted into green space, with people offered an opportunity to relocate to higher ground.

Ahmed El-Adawy, an environmental researcher who investigated the failure of the Derna dams, said: 'It's really hard to only ask an engineer. What's needed is a social study of what's there: who owns the land, what's the price of the land? ... And then you can give protection for this land use.'[16]

As the effects of climate change have become ever more visible, with increasingly destructive floods, heatwaves and wildfires, it's become harder to deny that the planetary climate is changing. But the retreat of democracy in the world's most fragile states and the rise of the populist right in the rich world are frustrating efforts to deal with both the causes and consequences of this shift.

Political leaders whose appeal lies in showing strength through a crisis may see little benefit in protecting their people from shocks. That's most apparent in the richest country in the world, where populism is in the ascendant and climate-scepticism is now embedded in the government.

CHAPTER 9

HURRICANE SEASON

Michael Bobbitt would not describe himself as a risk-taker, though it's a phrase others might use of a man who finds storms exhilarating.[1] He never feels more alive than when there's a monster hurricane bearing down on Cedar Key, an island town so remote and – usually – so tranquil that it doesn't have traffic lights. Cedar Key is on the island of Way Key, linked by bridges across salt marsh and shallow water channels to the Big Bend region of Florida, the curving stretch of coastline where the state shifts from the north-western 'panhandle' to the peninsula that hangs south down to Miami and the Everglades. It's a sparsely populated corridor of small towns that rely on fishing and tourism.

Bobbitt is more likely to describe himself as foolhardy, while many of his neighbours are best characterised, he says, as stubborn. These are people who have seen plenty of wild weather blow in from the Gulf of Mexico to batter the Florida coast and are happy to hunker down in the face of a storm. Climate science is a polarising topic on the Gulf Coast, with people belonging to one tribe or the other. But while

Bobbitt is a southern American man, a military veteran and a business owner – typically a member of the tribe that is sceptical on climate change – he has no doubt the water is getting warmer and it seems no coincidence that there are more ferocious storms slamming into the coast every year. Bobbitt, a playwright and novelist, has a side line as a clam farmer and the warmer seas have been deadly to his shellfish.

The US is skewed to its coastline, where more than 40 per cent of the population lives and wealth is concentrated. More than 800,000 people have moved to the coast every year, for the last five decades, lured by work in tourism, the oil and gas industry and shipping, as well as the desire to live by the beach. They are people like Bobbitt, who grew up in Florida's interior and discovered Cedar Key when he was training to be a pilot (it has the appeal, to an adrenaline addict, of an exceptionally short runway).

It's become commonplace to talk of a future in which humanity will move to escape stretches of uninhabitable terrain, but the reality of the last few decades appears just the opposite. Around the world, the number of people living by the sea is booming, increasing by nearly 30 per cent between 2000 and 2018. Vast numbers of people in the world's richest country are moving towards danger; in the last few decades, the population of the US regions that are most vulnerable to hurricanes has grown to 60 million and it has increased every year with just one exception: 2005, the year Katrina flooded New Orleans.

In that time, the Gulf and Atlantic coastline of the US has been hit by some of the most damaging hurricanes on record.

In the last 25 years, there have been 15 hurricanes that, individually, caused more than $10 billion worth of damage.[2] Katrina and Helene, two disasters nearly two decades apart, show how climate change is transforming life in the US – and how the country is failing to adjust.

Helene began as a broad low pressure system in the western Caribbean in late September 2024, the winds blowing towards its centre and rising where they met. It consolidated into a storm and then a hurricane, intensifying as it crossed the very warm waters of the Gulf and accelerating north towards the Florida coast. By 26 September, just as Helene was about to make landfall, a US weather satellite picked up frequent lightning flashes in the eyewall, the ring of heavy rain and powerful winds surrounding the hurricane's centre. In the satellite image, the eyewall seems to glitter with lightning, a sign that its wind speeds were picking up rapidly.

Officials had given evacuation orders, but in this part of the world, said Bobbitt, these are often treated as a 'suggestion'. One of the reasons for their calm was the fact that it was a little more than a year since a powerful hurricane last ripped through, when Idalia hit the Gulf Coast in August 2023, flooding homes and businesses but taking relatively few lives.

This time, while most people took the order seriously, around 50 out of a population of 750 on Cedar Key stuck around to witness Helene. Bobbitt was expecting it to be far worse, a hurricane that would make Idalia look like an 'afternoon thunderstorm', but, still, he was one of those who stayed. His house is up on a hill, which he judged high enough to escape the storm and he reckoned he could help elderly neighbours who were too stubborn to evacuate.

The colossal storm made landfall late on Thursday 26 September as a Category 4 hurricane. It had peak wind speeds of 225 km/h, which is strong enough to snap trees and rip the roofs off houses. Helene swept seawater onto the land, generating a 4.5-metre-high storm surge, but because the storm was moving rapidly, it shed very little rain in Florida. Bobbitt recalls a deafening sound of wind 'like a freight train' and the sea rolling into his town. People who live by the Gulf are not used to seeing much in the way of waves; its water is usually calm and clear. But, that night, the sea churned over the land like a washing machine, with waves sweeping through the town.

Bobbitt heard water crashing onto houses, followed by crumpling sounds as buildings gave way. Leaving the safety of his house to get a better look at what was happening to the town, he quickly found himself up to his neck in water that shoved him along with the force of a river in full fury. He grabbed hold of a stop sign's metal pole, clinging on, as a car and then part of a building washed by him. Knowing it was causing devastation, Bobbitt was still enraptured by the hurricane, a glimpse of an eternal and irresistible force. The sea is the great provider, he thought at a calmer moment, but an 'angry god' – a force that gives life constantly, sometimes turning on its worshippers and taking everything from them.

At sunrise, Bobbitt saw that entire rows of houses had been swept away completely, while the post office, food store and many restaurants were shattered. But the true disaster still lay ahead.

Helene was unusually strong and, at more than 600 kilometres wide, it was also one of the biggest storms ever to hit the Gulf Coast. Storms usually lose their strength after making landfall, as they can no longer draw heat and moisture from the ocean, but Helene was different, because it was so big and moving so fast that its winds were still powerful as it headed inland.

In late September, the town of Asheville, North Carolina, often enjoys crisp sunny days with clear blue skies. Tourists gather to drink craft beer in the converted warehouses and workshops of the River Arts District, a former industrial zone facing the French Broad river. In the mountain forests above the city, the leaves put on a dazzling display of red and gold.

A North Carolina state senator once referred to Asheville as a 'cesspool of sin' and some of its residents took pride in that statement.[3] It's a liberal dot in a conservative state, with a large LGBT community and many artists and musicians. J.T. La Bruyere, who owns and manages two coffee shops in the city, one in the River Arts District and one downtown, said members of his family had moved there from New Orleans, partly to escape Louisiana's hurricanes and partly because Asheville was 'just weird enough to scratch some itch' of the city they had left behind.[4]

In the autumn of 2024, it had been raining hard in Asheville and, on the Thursday, in the hours before Helene made landfall, business owners in the River Arts District started to worry about the water. The French Broad, which flows north from the Appalachian mountains, had flooded before, but they expected, at worst, a little water to soak the floors of their buildings. Lauren Turpin, co-owner of Pleb Urban Winery, said: 'When we built out there, we knew it was a floodplain. We had done our homework.'[5] The winery's tanks and barrels were already raised off the ground, but she decided not to move anything out of the building.

At the Grail Moviehouse, an indie cinema, co-owner Steve White put sandbags by the back door, covered the projectors with blankets and moved the candy and popcorn. At 12 Bones, a barbecue restaurant, owners Angela Koh and Bryan King told staff to close early.

The cinema was screening the horror movie *The Substance*. 'If we could choose a metaphorical movie to have on screen,' said White, 'that was the one. It was about transformation, something tragic, and it's gross – everything that happened to us days later.'

By the time Helene rolled up into the mountains, the soil was waterlogged and both the French Broad and it's tributary the Swannanoa were high. When the storm hit, the rivers burst their banks and Asheville, so far from the sea that at least one real estate broker had pitched it as a 'climate haven', was inundated.

Roads were submerged, the bars and art galleries of the River Arts District deluged in mud and downtown streets strewn with debris. Clean water was cut off to the city, power lines were brought down and the cellphone signal blinked out. Landslides cut off routes out of town.

When Koh and her husband got back to their restaurant, they found chairs that had been shoved up to the ceiling by the flood and left dangling from the rafters. Pleb Urban Winery had disappeared, with only the foundations left to show where a building had been. Five metres of river water had rolled into the Grail Moviehouse and lifted part of the roof off. 'It was so much worse than what anybody thought was possible in the mountains,' Koh recalled. 'And it happened so fast. It wasn't like over the course of a week. It was literally 36 hours.'

Remarkably, down in Cedar Key, there were no reports of any deaths. While Helene's storm surge was deadly in other parts of Florida, the rain that the hurricane carried into North Carolina claimed many lives, including a seven-year-old boy, Micah Drye, who drowned along with his grandparents when their house in Asheville collapsed. In all, Helene caused 250 fatalities, making it the deadliest storm to strike the mainland of the US since Katrina, which is reckoned to have claimed around 1,400 lives.

Helene's destructive sweep from the coast to the mountains underlined one of the ways in which climate change is transforming the dynamics of hurricanes. As surface water temperatures in the Gulf of Mexico reach exceptional levels, hurricanes' wind speeds are increasing dramatically in the hours before they make landfall; Helene went from 130 km/h to 225 km/h in the space of 24 hours. These rapid gear shifts give people far less time to react and officials in charge of emergency services a narrower window in which to decide whether a storm is likely to inflict a glancing blow or a direct hit that requires an evacuation.

The storm knocked out Asheville's power supply and wrecked part of the town's water system. For weeks, there was no running water. Whether it was the brush with mortality or just the loss of utilities, a mood of living in the moment seized many in the town: defunct freezers were emptied of steaks, with neighbours invited over for barbecues on the gas grill. Water from hot tubs was used to flush toilets. Rich societies are increasingly geared towards the consumer while populists emphasise the tribe, but it was the strength of the community's bonds that helped Asheville cope with the aftermath of disaster. White said: 'Asheville is very much a haven of enlightenment. We like to think we take care of each other.'

The worst of the damage in Asheville had been confined to streets by the rivers. Many homes in the city had power back within days, although it took nearly eight weeks for potable running water to be restored. Roads were cleared of debris and the National Guard delivered food, water and other supplies, while aid group World Central Kitchen brought its field kitchen: an all-terrain vehicle serving sandwiches and hot meals. Restaurants opened up as soon as they could and started providing food as cheaply as possible. 'By donation or for five dollars,' White recalled, 'you could get giant meals.'

Koh and King had a second branch of 12 Bones around 16 kilometres south of Asheville; here there were downed trees and no power, but little other damage. At his downtown coffee store, La Bruyere got back up and running. It was a different story in the River Arts District. 'When we finally went back down,' White said, 'we were trying to pick through the building to see if there was anything worth saving, walking around with other business owners doing the exact same thing.' White picked up a few tokens of the past, such as the cinema sign that had been painted by a customer. Then, he thought, this chapter is over.

The Federal Emergency Management Agency, established in the 1970s, coordinates the national response to a disaster in the US. While states and cities play a front-line role, they frequently call on FEMA's help when it appears they will be overwhelmed. FEMA musters supplies, organises relief efforts across state borders and provides financial aid, which includes reimbursing state and local governments for the costs they incur.

In the aftermath of Helene, some residents complained FEMA was slow to arrive. The agency said that rescue teams were on the ground in the immediate aftermath, while trucks and planes shipped in supplies. But the rugged mountain landscape, with roads choked by fallen trees, made it hard to reach those in need.

Some took matters into their own hands. A group of military veterans organised their own relief efforts out of a Harley-Davidson dealership in Asheville. The volunteers used a fleet of privately owned helicopters to deliver food and fuel, paid for by donations, in a rescue mission nicknamed the 'Redneck Air Force'.[6] A private military contractor led mules carrying packs of food, water and nappies up mountainsides that were now impassable to vehicles. Governments can act on a far bigger scale, but the volunteer efforts strengthened a sense of self-reliance – and, for some, heightened distrust of the state.

Conspiracy theories sprang up like weeds in the aftermath. Marjorie Taylor Greene, a far-right Republican congress-woman, posted on X: 'Yes, they can control the weather.' One theory which swirled around the North Carolina village of Chimney Rock, hit by mudslides and flash floods, was debunked in early October in a lengthy statement from Chuck Edwards, a Republican congressman in North Carolina.[7]

Hurricane Helene was not geoengineered by the government to seize and access lithium deposits in Chimney Rock, Edwards wrote. Local officials were not abandoning search and rescue to bulldoze over Chimney Rock. Other rumours attacked FEMA and he addressed those too. FEMA has not diverted disaster response funding to the border or foreign aid.

A few weeks later, Donald Trump visited North Carolina and gave a press conference amid an apocalyptic landscape of storm debris. With dust on his dress shoes, the man running for president in the 2024 elections repeated the false claim that the emergency response had been hampered because FEMA had diverted funds to shelter undocumented migrants. Edwards stood quietly behind him as he spoke.

Sea levels around the mainland US are rising faster than the global average, gaining around 15 centimetres over the last three decades. The rising seas mean there will be deeper and more frequent coastal flooding over a wider geographic area. Higher storm surges will bring the risk of further devastation to coastal cities.

The US confronts difficult choices about where to defend, and where to retreat. These choices will reflect the legacy of previous disasters and how the country responded to those, but they will be haunted, above all, by the ghost of Katrina.

When the French founded Nouvelle-Orléans on the banks of the Mississippi in 1718, it was, despite hurricanes and mosquitoes, an ideal spot for commerce and settlement. Between Lake Pontchartrain, which lies north of the present-day city, and the river there is a ridge of elevated land known to the native Americans as Balbancha, the 'place of languages'. Men paddling canoes along the lake's shoreline could unload cargo here and hike up to the river, making it a place where communities met to trade, filling the air with the babble of their different tongues.

Favoured by its geography, New Orleans prospered. The great river was a natural conveyor belt for goods from further north in the continent to New Orleans, just 160 kilometres from the sea. The river's deposits of silt and the subtropical climate created fertile conditions for crops of sugarcane and cotton.

Wealth decided how far above sea level people lived. The French Quarter, the grid of streets which is the city's oldest neighbourhood, lies behind a natural levee, a spot where the Mississippi piles up a bank of sediment that shelters the land beyond against flooding. This is now Decatur Street, originally Rue de la Levée, a stretch of jazz bars and restaurants. The old heart of the city, the 'sliver by the river' as it's known, is ironically both the closest to water and the safest from flooding.

The French extended the city further into the swampy river delta and, after buying the Louisiana Territory for $15 million in 1803, the US government did the same. Enslaved Africans, along with prisoners and immigrant labourers, built artificial levees, earthen embankments that held the water back. Taming the river behind man-made levees meant that it no longer dumped sediment on its banks and, without those deposits, parts of New Orleans gradually sank further down into the swampland.

Water surrounded the city. The Mississippi wound in a crescent through its heart, before flowing south to empty into the Gulf of Mexico. Lake Pontchartrain, which is not a true lake but an estuary connected to the Gulf, limited the city's expansion to the north. To the east, west and south of the city were coastal wetlands that acted as natural buffers to floods. These were drained, both to eliminate mosquito breeding grounds and extend the city.

By 1840, New Orleans was the third biggest city in the country, after New York and Baltimore, growing fat both on the trade in cotton and on a darker business; the city was the largest market for enslaved people in the US.[8] While the construction of railways and roads reduced its importance as a port, it remained the biggest city in the American South until the middle of the 20th century.

Around the world, poorer communities are often forced to live on marshy ground, a history preserved in place-names like Bogside in Londonderry, a Catholic neighbourhood of a Protestant city, or Moss Side in Manchester, where many Afro-Caribbean immigrants settled. In New Orleans, this was the Lower Ninth Ward, an area of marshy land east of the city. It was settled by European immigrants – Germans and

Irish – and by African-Americans freed after the Civil War. The Industrial Canal, built in the 1920s to link the river to Lake Pontchartrain, cuts the Lower Ninth off from the rest of New Orleans.

Black people had been explicitly restricted from owning property in some neighbourhoods of New Orleans by clauses in property deeds. Such racially restrictive covenants were common across the US; although the Supreme Court ruled in 1948 that they could not be enforced by the courts, the words lingered on legal documents and gave an indication of the kind of person who was welcome in a neighbourhood. A covenant in the Lakeview neighbourhood of New Orleans stipulated that 'no lots are to be sold to negroes or colored people'.[9]

The Lower Ninth was different. Less than half of households in New Orleans owned their own homes, but the Lower Ninth had high rates of home ownership, with many black households owning homes free of mortgages that had been passed down through generations. In some cases, these dated back to men and women who had been freed from slavery. Such intergenerational black wealth remains rare in the US.

The fact that there was no rent to pay helped people survive in a city where many earned low wages. It was a good place to raise a family too, with plenty of space for children to play under the shade of trees. There were backyards large enough to keep chickens and even horses, although it was sufficiently close to the city to get a bus to work downtown.

At its lowest point, the Lower Ninth is 1.2 metres below sea level. The difference in elevation between the highest and lowest parts of New Orleans is not vast – the highest parts of the French Quarter are just 3 metres above sea level – but, in a flood, a few centimetres can make a difference.

In September 1965, Hurricane Betsy swept a storm surge from Lake Pontchartrain into the Industrial Canal that breached the canal wall and inundated the Lower Ninth. As many as 81 people died and a trail of devastation was left across the eastern side of the city.

The next day, President Lyndon Johnson got a call from Russell Long, a Louisiana senator and a political ally. The senator told him that a 400-year-old tree had toppled on top of his own home, although his wife and children were safe. Long told the president 'that Hurricane Betsy picked the lake up and put it inside New Orleans'.[10] The senator went on: 'Mr President, we have really had it down there and we need your help.' Long combined his cajoling of the president with an urge to calculate the political advantage he would win from making an appearance at the disaster scene.

The president flew there that evening on Air Force One. From the air, they could see water covering the city, in some places up to the eaves of houses. His motorcade drove out to a bridge and Johnson got out to look down at the water engulfing the neighbourhoods beneath. People who had been rescued from their rooftops were walking to safety across the bridge. Johnson stopped at a school that had been turned into a refuge. It was a 'mass of human suffering', according to the official White House diary entry, with hungry and thirsty people crying out constantly for water.

Alongside the death and distress it had brought to New Orleans, the extensive property damage made 'Billion-Dollar

Betsy' the costliest storm the US had ever seen. The next month, Congress authorised a flood protection plan for New Orleans. The US Army Corps of Engineers took charge of building a vast system of levees and floodwalls around the city. But the ring of armour had gaps in it.

By 2005, costs had reached nine times the original estimate and the work was still not complete. Some parts of the defences were left unfinished due to lack of funds. The wall by a pump station on a canal that drained rainwater from the city out into Lake Pontchartrain was lower than its adjacent floodwalls, providing a route for water to flow into the city.

Flood defences known as I-walls, shaped like a capital I, provided protection along the canals. These are vertical columns of concrete buried into the top of a levee. The alternative is to build a 'T-wall', which is shaped like a letter T with the horizontal part lying buried in the levee. This is a much more expensive but also much sturdier option.

Steel sheet piling was driven into the earth below the I-walls, providing a watertight barrier below the ground. In the 1980s, the Corps decided that the original design criteria for the sheet piling was too conservative, going deeper into the ground than necessary. The army engineers issued revised guidance that allowed the steel sheets to be planted less deeply, saving money.

The official inquiry report into what went wrong makes only a subtle reference to this, but the political climate had shifted in the 1980s.[11] Ronald Reagan had won the 1980 US presidential election determined to shrink the gap between federal spending and revenue, and to do so by cutting government outlay. The amount of federal funds available for water development projects did not increase after 1980, the report says, 'even as demands for civil works funding increased across the nation'. The direct cause of the Katrina disaster was a failure of the defensive system around the city. Behind that lay decisions based on saving dollars.

Hurricane Katrina started life out in the Atlantic. On Sunday 28 August 2005, its winds blew at about 280 km/ hour, making it a category 5 hurricane – the most severe. By the time it had made landfall in Louisiana around 6 a.m. the following morning, it had weakened to a category 3. Hurricanes are ranked one to five on a scale created by a structural engineer, Herbert Saffir, and developed by a meteorologist Robert Simpson. It's based on wind speed.

The Saffir–Simpson scale offers a reliable guide to the damage a storm's winds will inflict, but offers much less predictive power when it comes to a storm surge and the height of waves, because they are determined not just by the wind but the shape of the coastline and the seabed as a storm approaches land. A wide and shallow continental shelf, like the Gulf Coast, enables a higher storm surge.

Katrina's winds whipped up surges on Lake Borgne, an inlet of the Gulf of Mexico to the east of the city, and on Lake Pontchartrain north of New Orleans. The surges swept against the shoreline levees and funnelled into the confined spaces of the city's canals. It is hard to be sure precisely how powerful Katrina's surges were because every tidal gauge was destroyed in the storm, as were many buildings that might have carried evidence in the form of a high water mark. But the data that exists indicates the surge that struck eastern New Orleans was more than 5.5 metres high.

Wetlands are likely to have absorbed some of the impact of the storm in its early stages, providing some protection to parts of the city, but Katrina was so massive, driving such a vast surge, that it had saturated the marshes even before the peak of the storm arrived.

The floodwalls along the Industrial Canal breached in two places, inundating the Lower Ninth Ward. This happened in the early hours of 29 August, with the first breach taking place before the storm had even made landfall – a section of 'I-wall' was tilted far enough to open a crack that let water under high pressure reach the clay at its base. With an immense force of water brought to bear on this clay, the levee slid backwards, letting the flood pour through the gap and into the streets beyond.

The storm breached the city's defences in 50 places and floodwater occupied eight-tenths of the city.[12] Buildings splintered while others were wrenched from their foundations and shoved along the street. Hundreds drowned or were struck by debris carried along in the raging water.

Before Katrina, around 485,000 people lived in New Orleans. Around 150,000 stayed to ride out the storm; some did not leave because previous storms had spared the city and an evacuation might have been a waste of their time. Others lacked their own cars or had responsibility for elderly relatives. After the flood, many survivors converged on the Superdome, where power had been lost on the day Katrina hit and where there was no light or air conditioning to temper the muggy heat of a Louisiana summer. Officials struggled to provide the people sheltering there with adequate supplies of food and water.

New Orleans was a singular disaster in US history. On a smaller scale, the town of Indianola in Texas had been wiped out by a storm surge in the 19th century and was never repopulated, but this near-total inundation of a major city had never happened before.

Remarkably, the first reaction of some officials was that it had been a near-miss, that the city had 'dodged a bullet'.[13] This was true of neighbourhoods on higher ground, but not for the low-lying districts of lower-income households.

FEMA struggled to deal with the scale of the disaster, which overwhelmed its systems for delivering supplies, leaving survivors stranded without food and fresh water for days.[14] A Congressional investigation later found that the agency had lost a number of disaster specialists and senior leaders in the years leading up to Katrina.

For months, many districts had no running water or electricity; it was more than a year before street lights were back in the Lower Ninth. In parts of the ward, the concrete steps that led up to house entrances were all that was left of the houses; stairs to nowhere.

Katrina demonstrated the widespread and lasting destruction that can be inflicted by a storm surge, flattening housing over a large area and crippling infrastructure. It was the costliest storm in US history, causing $201 billion of damage, compared with the $79 billion bill for Helene.

In the immediate aftermath of the storm, a different future was sketched out for New Orleans. A German journalist found Finis Shelnutt, a real estate broker, drinking champagne in the French Quarter and anticipating the opportunities the disaster opened up.[15] Shelnutt told the reporter: 'Most importantly the hurricane drove poor people and criminals out of the city, and we hope they don't come back.'

A few weeks after the hurricane, real estate developer Joe Canizaro mused about the city's run-down housing and poorly performing schools. 'I think we have a clean sheet to start again,' he suggested. 'And with that clean sheet we have some very big opportunities.' The mayor established a Bring New Orleans Back Commission and Canizaro headed that commission's urban planning committee.

On 11 January, the local newspaper, the *Times-Picayune*, published a front-page illustration based on the final report of the mayor's commission. Green dots covered areas designated for parkland, including swathes of the Lower Ninth. There's a more flattering interpretation of the Green Dot plan; that it emphasised rebuilding on higher ground while establishing more green space in areas that could soak up floodwater.

But it was regarded as a top-down project to transform the city, one in which the racial hostility of its historical zoning had become evident once again. The idea that a flood would drive people out of the Lower Ninth was hard to reckon with. After all, as Rashida Ferdinand, an artist living in the ward, put it: 'People have lived in this bayou marsh area for centuries.'[16]

The suspicion stirred up by the green dots was strengthened by the government's Road Home programme, which distributed grants to people whose homes were destroyed by Katrina. The grants were paid out on the basis of the pre-hurricane value of a claimant's home, or the cost of rebuilding, whichever was lower. This was, on the face of it, fair – but, of course, black homeowners in the Lower Ninth lived in homes of lower value than white neighbours on high ground, even if their rebuilding costs were similar or higher.

Five black homeowners brought a legal challenge and a federal judge agreed it was 'regrettable' that the rebuilding of the city appeared to have disadvantaged African-Americans.[17] The homeowners who brought the case ended up settling out of court with federal officials. Regrettable or not, the character of the city was changing. Shelnutt's prediction that the storm would expel the poor was borne out.

While many in the Lower Ninth had been homeowners, few were insured and the modest grants available under Road Home meant few could afford to return. The population of the Lower Ninth tumbled from around 14,000 before Katrina to 4,000 after it. The district was redeveloped in pieces, with weeds growing thick on lots that were left derelict for years. Within a few years, New Orleans had one of the highest levels of vacant and run-down properties anywhere in the country.

The New Orleans flood protection system was redesigned and rebuilt by the US Army Corps of Engineers, who turned to a Dutch engineering firm for advice. I-walls were replaced with sturdier T-walls, while the slopes of levees were 'armoured' with grass that would hold the soil in place. The Dutch advised restoring marshland around the city to provide natural defences – working with nature, rather than against it.

In 2007, the state government announced plans to invest billions to revive a vast complex of wetlands south of New Orleans, in the largest ecosystem restoration in US history. The flood protection system was rebuilt at a cost of $14.5 billion and held through subsequent hurricane seasons.

Behind its new walls, the character of New Orleans changed. African-Americans once made up 67 per cent of the city's population, but that share dropped to 60 per cent after the storm. Overall the city, while still smaller than before the hurricane, is growing again. It has been revived partly by an influx of Latino immigrants, but also by university students and middle-class professionals, who have settled in the higher terrain of the city's older quarters.

Twelve years after Katrina, an area of deserted land on the northern edge of the Lower Ninth was turned into a wetland in an initiative led by Ferdinand, whose family had lived in the neighbourhood for generations. Unlike the green dot map, it did not come as a surprise; early on, a series of public meetings let people share their views on the design.

A derelict zone that had been used for illegal dumping became a wetland park and nature trail, a swathe of green that was soon home to beavers, otters and dozens of species of wild birds.

On the face of it, the resurrection of New Orleans looked like a triumph, at least for some. The city was wrapped in concrete and steel, and came back to life behind its walls. But nearly every decision that was taken spells trouble for the future. New Orleans is no longer the third biggest city in the country; it's not even the biggest city in the southern US (that's Houston, Texas). But billions of dollars have been spent on defending it.

When they were built, the flood defences provided a '100 year' level of protection – able to withstand a storm surge that had a 1 per cent chance of occurring in any given year. But Louisiana is sinking – partly because of the extraction of fossil fuels – and sea levels are rising, which means that this level of protection is no longer guaranteed.

In 2025, the plan to restore coastal wetlands was cancelled by the governor, Jeff Landry, who was concerned that it was no longer financially viable and threatened the livelihoods of fishermen. The challenge facing the city now is whether millions more will need to be spent on the never-ending task of further raising its engineered flood defences.

Across the country, neighbourhoods and entire towns must grapple with the need to relocate. But the example of New Orleans was of a piecemeal relocation, skewed on grounds of race, that happened after the crisis rather than being planned ahead of time – the worst possible way to deal with the changing reality of the planet.

Cities like Asheville, where the population is predominantly white and more than half are college-educated, tend to rebound and accumulate wealth after a disaster, research indicates. Black and less well-educated communities tend to lose wealth after suffering damage in a natural disaster.

The past and present of racial injustice in the US helps explain why.

From New York to Houston, the big cities along the Atlantic and Gulf coasts lack the extensive fortification that now protects New Orleans. Wetlands can adapt to rising seas by retreating inland, but their path of retreat is increasingly closed off as more people move to live near the sea and the zone of human settlement spreads out. It's a phenomenon known as 'coastal squeeze', in which cities eliminate their own natural defences.

In July 2025, a tropical storm knocked out the pumping station in the town of Hillsborough in North Carolina, sending thousands of cubic metres of dirty water into the rivers. The pumping station, built in the 1970s near the Eno River, is regularly flooded in severe weather, but a few years earlier, the town had received a federal grant of nearly $6 million to relocate the station to higher ground. Work was due to begin at the end of the year.

The local government in Scranton, a city in north-eastern Pennsylvania, planned to buy and demolish 18 homes in neighbourhoods with recurrent damage from stormwater and restore a natural floodplain. Officials in Hoquiam, a city on a coastal estuary in Washington state, on the US Pacific coast, proposed constructing a 10-kilometre levee to provide flood protection.

Central government funding for all of these projects, along with hundreds of others, was withdrawn after the Trump administration cancelled a programme called Building

Resilient Infrastructure and Communities. In a statement, FEMA, now under new management, described the spending as 'wasteful'. Yet the programme cost $4.5 billion, a tiny fraction of the damage an individual hurricane can cause – and it was aimed at preparing for disaster rather than cleaning up after it. The decision to withdraw funds from projects that were already underway seemed both chaotic and reckless in its willingness to put people and property in harm's way.

On the 20th anniversary of Katrina, in August 2025, employees at FEMA wrote to Congress to remind them that the hurricane that ravaged New Orleans was not just a natural disaster but a man-made one.[18]

The letter claimed that the Trump administration was choking off FEMA's funding, diverting its employees to immigration enforcement, ending a grant programme to help communities prepare for disaster and ignoring the facts about climate science. Its acting head David Richardson, appointed in May 2025, had startled colleagues by saying he did not know the US had a hurricane season. The government department that oversees the agency said this was a joke.

If that was an attempt at humour, it seems benign compared to what's now the official view of climate change. In July 2025, the US Department of Energy published a report that accepted the planet is warming, but said that carbon dioxide-induced warming 'appears to be less damaging economically than commonly believed'.

More than 85 climate scientists criticised the publication, calling it biased and full of errors.[19] 'Their report seeks to downplay the risks of record-breaking heat, intense rainfall, worsening wildfires, rising sea levels and widespread health harms – all well-established by decades of peer reviewed science,' the scientists wrote.

Few wealthy countries now look quite as vulnerable to disaster as the US, with its main disaster relief agency ill-prepared for catastrophe and its government increasingly unwilling to listen to scientists.

Months after the River Arts District in Asheville was submerged, its recovery was slow. The owners of the Grail Moviehouse were looking at a new beginning in a factory building 16 kilometres away. The owners of 12 Bones focused on their second location, having decided it made little sense to go back to the riverside. After the storm, Lauren Turpin wanted to keep making wine, though it is likely to be a hobby after her investment was washed away. One of the reasons she wanted to stay in wine-making was its connection to the environment – how vines are affected by wind, rain, humidity and soil.

For now, there are more questions than answers. 'This is going to happen again,' Turpin said. 'Do we even develop in those areas? And if we do develop, how do we develop sustainably so it doesn't cause someone like me to lose their life savings?'

CHAPTER 10

MONEY

When there's a big storm, especially one of those that seems to blow up out of nowhere, the minds of Chris Hook and his wife spool back to the day when a summer tempest sent rain chasing across their garden, down three steps and sweeping into their basement bedroom.[1] There's another way the flood lives on, aside from this mental glitch. Their annual insurance premium rose tenfold to £3,200 in the space of a few years.

It's not an impossible sum for a middle-class couple, but it's one indication that the impact of climate change on the property market isn't some far-off danger. It's here. Chris and his family live in the ground floor and basement of an Edwardian house – the kind with high ceilings, a solid brick exterior and plenty of space for their twin boys to play, including a tidy yard paved with a lick of artificial grass. It's high on a hill in north London, far from the Thames and down a gentle slope from Alexandra Palace, one of the highest points in the city. It's one of the last places anyone would expect a flood.

The damage was inflicted by a July thunderstorm in 2021, in a work-from-home year when Chris sat at his laptop in the basement. 'All of a sudden, it started to rain incredibly hard,' he recalled. 'It got very dark.' The rain was weird, he remembers, moving at impossible, nightmarish speed, pooling in a side return outside their basement.

Within seconds, what began as a trickle on the ground became a body of water pressed against the external door of the basement bedroom. He tried to bail the water out of the side return, tipping bucketfuls down the drain, but that had no impact at all, apart from getting him drenched, clothes clinging to his skin. Then water swept into the basement. He started moving valuables out of their bedroom: jewellery handed down from grandparents, books the couple had read to their boys as toddlers. But, within half an hour, the flood was up to his chest. Rescuing anything more was hopeless.

The damage was not structural, but it was significant enough; plasterboard and electrical wiring needed to be replaced, along with the washing machine and tumble dryer. Clothes and shoes were ruined, but nothing of sentimental value had been lost. The couple's insurance paid out, covering a £45,000 bill – most of that for the rebuilding work. Then, it got complicated.

The couple's original insurer pulled out of home insurance and sold their portfolio to another company, which declined to take the risk on the Hooks' home. Just one insurer would accept their business and the couple were charged more than ten times their original premium. Floods don't just cause physical destruction and social upheaval; they have financial consequences too. As rainfall becomes more intense and sea levels rise, parts of the world may become uninsurable before they are uninhabitable.

And because so much of the world's economy is tied up in property wealth, that will echo through the entire financial system.

All of modern history can be described as the history of managing risks. There are the risks that humanity has created – the danger of runaway artificial intelligence or ever more powerful weapons – and the risks we have magnified by living packed together in ever-bigger cities, such as viruses, fires and floods. Disaster insurance holds a special place in that history: it is the way that we manage the financial shock that follows a hurricane ripping off a city's roofs or a storm surge smashing into a coastal town. And it was born in fire.

More precisely, it was born in the spectacular fire that broke out after the long, drought-stricken English summer of 1666 and consumed London, after which a few enterprising men spied opportunity in the smoke and ruin, creating the world's first fire insurance companies. Businesses have always tried to mould their customers, but insurers have done this more than most, attempting to shape society to reduce and manage risk so that it can be kept within comfortable (and profitable) boundaries. The fire insurers who went into business after the Great Fire of London did not just earn money from fees and agree to reimburse losses. They hired their own 'watermen' to put out blazes, the forerunners of modern fire brigades, and created 'fire mark' plaques for houses to show which ones were insured.[2]

The trouble with floods is that it's easier to insure against a risk that's scattered and manifests itself at random across a population. Floods are the opposite, especially storm surges. Their impact is catastrophic, highly concentrated and they often strike the same places again and again: low-lying coastal regions like Louisiana or steep river valleys. Or places that combine both geographies like Derna in Libya.

Compounding that difficulty is the fact that there's now much more life and wealth in the path of floods than ever before. After the 1953 floods, the belief that sea walls and surge barriers offer enduring protection has encouraged many more people to settle along Britain's coastline. This is also true everywhere in the world. The human cost of a flood is clear to us from disasters like Katrina, but the danger now is that floods could also deliver a wider financial shock. Powerful and sustained damage from floods could result in more than just a local or regional tragedy, starting a shock-wave that ripples through the wider economy, beginning with the most valuable asset in most people's lives: their home.

Through history, financial crises tend to evolve in three stages, whether they involve tulip bulbs, railway stocks or houses. The first is a cautious stage, when lenders place restrictions on loans. The second is a speculative stage, as buyers become greedier. The third is one of excess, when the prices of tulip bulbs or railway stock are rising rapidly and every buyer hopes that they can sell for a profit to the next one: the 'greater fool' stage of the cycle. What often happens next is a sudden loss of faith in the system and a panicked attempt to sell assets – and then a crash.

The phases of the cycle are obvious with hindsight, but when they are happening, it's hard not to get carried along in the excitement. None of the people buying tulips or railway stocks feel as though they are making a mistake. After all, everyone else is making money. As a banker put it before the financial crisis of 2008, 'as long as the music is playing, you've got to get up and dance'. As that banker acknowledged at the time, the tricky part is what happens when the music stops.

Protecting homes from flood risk is a dance between business and government. It's usually businesses that provide mortgages and manage the risk through insurance, while the job of the state is to provide flood protection. Both businesses and governments play some part in helping people adapt their homes and their lifestyles: elevating plug sockets and appliances or using steel and hard wood instead of fibreboard. Over the last two decades, as payouts for weather-related damage have soared to record levels, access to flood insurance in the most flood-prone parts of Britain has looked increasingly shaky.

A deal struck between the insurance industry and the government has provided a temporary solution, spreading the cost of flood insurance across the whole industry. Insurers pay a levy to fund the scheme, known as Flood Re, and in return they can pass on the flood risk part of a customer's policy. It was established by legislation in 2014 and was launched in April 2016, the understanding being that if the industry kept on providing home insurance for flood risk, the government would fund flood defences.

Nearly 350,000 households at the worst risk of flooding now have insurance policies underwritten by Flood Re, but if there was a 'gentlemen's agreement' between insurers and the state about flood protection, the government has struggled to maintain their end of it. Over the last few years, the UK has spent about £1.3 billion annually on flood defences, which is far too little.[3] The Dutch, with a coastline shorter than that of Essex, spend about two-thirds of this amount every year on reinforcing and maintaining their defences.

In Britain, after 14 years of governments devoted to cutting public spending, the construction of new flood defences lags behind schedule while existing defences have been allowed to crumble. In 2023, a UK government official acknowledged that a programme of flood defence intended to provide better protection for 336,000 homes would only cover 200,000, while thousands of homes were left at increased flood risk for lack of maintenance work.[4] Part of the reason was that building costs surged after the pandemic, but it was hard to shake off the distinct sense of a state failing to protect its citizens.

Flood Re does not cover houses built after 2009. This is intentional; it is meant to encourage house-builders to avoid locating new estates in the areas of highest risk. But it hasn't worked out like that. In recent years, a number of planning applications for large estates have gone against the Environment Agency's advice about flood risk. It can be a calculated gamble for a housing developer – build enough homes in a vulnerable place and the government may be forced to step in and build defences. Over the last decade, around 8 per cent of new homes in England have been built in areas at the highest risk of floods.

When houses on a new estate in Blyth, in the north-east of England, flooded twice in a year, the homeowners' insurance premiums quadrupled – and their new, more expensive policies did not include cover for flooding. Blyth is by the coast, but the water that invaded the new estate did not come from the sea but from two bouts of torrential rain in the spring and autumn of 2024. The people on the estate were first-time buyers, now heavily in debt for properties they can no longer insure and will struggle to sell.

At the other end of the scale, London's super-rich have created a subterranean shadow-city beneath the most opulent districts of the capital, with hundreds of basements excavated beneath mansions in Kensington, Chelsea and Westminster, many of them with swimming pools, cinemas, and saunas; one super-basement even has an artificial beach.[5] These luxury bunkers are a Bond villain's daydream while the sun shines, but a recipe for chaos in one of the flash floods now increasingly common in the capital.

Digging down removes space for water to drain, but it's just one factor driving increased vulnerability in cities. In a natural landscape, rainwater is slowed by greenery and seeps into the earth; in an artificial one, water moves rapidly over hard surfaces. In London, more than anywhere else in the country, front gardens have been paved over to provide parking spaces, allowing floodwater to rush across concrete before sweeping into rivers and canals.

From 'iceberg' super-basements in Kensington to homes hit

by repeat flooding in the north-east of England, there is little sign that either builders or house-buyers are conscious of the danger they face. The paradox of providing flood defence and guaranteeing insurance to houses at high risk is that it can remove the incentive to change, either by designing a home that's more resilient to floods or resettling in a safer place. And it may postpone a crisis in which some coastal homes cannot be insured or sold. Going back to the banker's quip, it keeps the music playing and the dancers swaying. But there is, with Flood Re, an agreed moment at which the music will stop.

The scheme was set up with a 25-year time limit, intended to give the insurance market enough time to prepare for change. But if nothing has changed by the time its provisions expire in 2039, if houses aren't being built differently and in locations at less risk of disaster, then home insurance in flood risk zones will likely become unavailable or be priced out of reach for many British households. The industry estimate is that around 2 million homes would no longer be able to access insurance for floods.

The damage caused by floods is so extensive that the costs of being uninsured can mount rapidly. When homes and businesses were inundated by rain in Doncaster, in south Yorkshire, in 2019, the estimated cost of losses at fewer than 200 uninsured properties was an average of £31,000 each.[6] That is, a bill of around £6 million for a single flood in one small city. Scale that up across the country and it's easy to see why some in the finance industry are growing nervous.

Nationwide, one of Britain's biggest mortgage lenders, was the first UK bank to say publicly that it has stopped granting mortgages on some properties with high flood risk.[7] But if insurers pull back, it will not be the last. Prospective buyers are likely to push for steep discounts on houses that are tricky to insure, creating neighbourhoods where prices could spiral downwards.

As housing prices in the UK have risen faster than wages, buyers are taking out longer-term mortgages. By the end of 2023, almost half of new mortgages in the UK were for 30 years or longer. The risk of extending mortgages this far into the future is that properties exposed to an increasing threat of flooding turn out to be worth less than expected, with implications for the banks that lent the money.

'Property is how people save for their kids,' said one bank executive.[8] 'If property becomes exposed to flooding risk, it could materially reduce the value. You thought it was worth £250k or 300k, but it could now be worth £150k.'

The Bank of England generally favours measured language, but its sober-suited men and women have blunt views on the potential risks.[9] In the most pessimistic climate scenarios, the Bank's policy wonks say, the 1 per cent of homes most exposed to flood risk could lose a fifth of their value. And if house prices drop and mortgage borrowers default on their payments, that will increase the losses that mortgage-lending high street banks suffer.

If a fresh crisis is brewing in the property sector, the population shift in the richest country in the world makes little sense. Still less, the fact that the ultra-rich remain willing to pay ever-higher prices for oceanfront property. In 2025, a mansion on Star Island, in Miami's Biscayne Bay, sold for

$120 million, a record even by the standards of one of the US's wealthiest cities.[10] Part of the answer is that the super-rich can afford to rebuild after a disaster, so, for them, the lure of a location still outweighs the risks.

But there's a deeper reason. Climate change may be a planetary shift, but its effects are local: small differences in elevation determine who gets flooded, while the shade of trees or the presence of green space makes a difference on days of extreme heat. Researchers at First Street, a US firm that models the financial risk of climate change, found that while the fire- and flood-prone Sun Belt states of California, Texas and Florida are still attracting newcomers, this pattern masked a more complex trend: the shift of population between neighbourhoods.

As its name suggests, Little Haiti is a predominantly black neighbourhood of Miami, a district of eye-catching murals, Caribbean groceries and cuisine.[11] It's also on a rocky outcrop about 3 metres above sea level, which has drawn buyers from Miami Beach, an oceanfront city about 8 kilometres away on a cluster of islands just offshore from mainland Florida. Miami Beach is wealthier than Little Haiti, but much more vulnerable to a rising Atlantic. Jeremy Porter, who leads research on the economic implications of climate change at First Street, said: 'Little Haiti has been a less desirable community historically, but there have been ten years of media coverage about persistent tidal flooding. People are asking their realtors about elevation. Elevation has become an amenity.'

A city like Miami is likely to grow, as its advantages outweigh the climate risk: a thriving economy, low taxes and beachfront living. It's smaller towns, in states like Minnesota that are already suffering from the departure of young people, where flooding is likely to accelerate decline.

In the US, the government takes an active role in managing prices in the insurance market. There are elected insurance commissioners, with the power to reject steep increases in premiums. In the year that Hurricane Helene hit Asheville, insurers in North Carolina pushed to raise rates in the state by more than 40 per cent. Public agitation about the proposed increase was clear; the state's insurance commissioner received thousands of emails, calls and letters. People told him they were already struggling with the cost of groceries and fuel. In the end, the increase was whittled down to 7.5 per cent.

This might be an expression of democracy, but the consequence is that it can be hard to put an accurate price on the degree of climate risk that some parts of the country face. The alternative to charging what it costs is that insurers either go bankrupt or withdraw. In Florida, more than 40 insurers have declared bankruptcy since 2003.

Unexpectedly, for a country associated with the risk-taking of free enterprise, much of the US's climate risk is in public hands. Across the country, a state-backed insurer protects those most at risk. The severe damage inflicted on Louisiana by 'Billion-Dollar Betsy' in 1965 and the withdrawal of private insurers from the most at-risk neighbourhoods led to the creation in the late 1960s of the National Flood Insurance Program. It now covers 4.7 million policies, nearly 90 per cent of the US flood insurance market.

In Florida, Citizens Property Insurance, which is owned by the state, was intended to provide wind-storm insurance to householders who could not get coverage on the private market. But since it was created in 2002, the insurer 'of last resort' has become the state's biggest provider after private insurers either went bankrupt or withdrew following a succession of costly hurricanes.

It's the same story in the mortgage market. Fannie Mae and Freddie Mac, two giant government-backed companies, buy up mortgages from banks and turn them into financial assets that can be traded. By doing this, they free up banks to lend more money to households and keep the whole system turning.

That means there's a significant amount of property value which is ultimately the responsibility of the government. The Congressional Budget Office, which produces non-partisan analysis, examined the risk for households with federally backed mortgages and found that, by 2050, the most vulnerable homes could suffer flood damage equivalent to 14 per cent of their value.

The danger is that, as the physical risk grows, the state's liability rapidly gets out of control, with bills mounting until governments struggle to function. At a local level, the obliteration of a town's business district by a flood could easily lead to economic collapse. That could have wider financial consequences, as a town that is unable to raise rates from businesses will struggle to service its loans. At national level, the collapse of a rich country like the US is unlikely, but escalating burdens will confront governments with a choice over just how much risk they are prepared to take on, on behalf of their most vulnerable citizens.

The answer, at least in the US, looks like it will be less and less. Unlike the UK's Flood Re scheme, which keeps prices affordable and spreads the risk across the whole insurance market, the US alternatives to the private market often cost more. Most American homeowners don't have flood insurance, but lenders require borrowers to take it out if they live in an area that's a high-risk flood zone.

The US's federal flood insurance programme is now in deep financial trouble. The NFIP gets its income from insurance premiums and, if there's a shortfall between that income and the amount it pays out, it's authorised to borrow money from the US Treasury.

After Katrina, the programme paid out more than $16 billion in flood-related claims and borrowed heavily to cover this record loss.[12] After Hurricane Sandy hit New York in 2012, it reached over $30 billion in debt, hitting its legal borrowing limit. At this point, Congress cancelled $16 billion of the debt so that the federal flood programme could borrow more. But by 2025, after several more storm surges and inland floods, the programme's borrowing had crossed $22.5 billion, leaving little space to cope with further disaster.

In 2022, the NFIP changed the way it calculated flood insurance premiums, moving to a system that more accurately reflected the risk an individual property faced. It now took into account not just a home's location within a flood zone, but included more variables, such as the distance to water. The change meant steep increases for some homes; the average premium in Plaquemines Parish, Louisiana, the marshy swathe of land around the final stretch of the Mississippi before it spills out into the Gulf of Mexico, soared from less than $700 a year to more than $8,000.[13]

For existing policyholders, the increases are capped at 18 per cent a year by law. But this is still a significant increase, especially for low-income families living in modest homes. The challenge comes when attempting to sell.

New buyers are taking this cost increase into account as they negotiate. 'People are negotiating down the price at the closing table,' said First Street's Jeremy Porter. 'There have been home devaluations because of insurance.' State-backed flood insurance may be expensive, but it protects millions of households from the financial risks of a flood disaster. It may soon be snuffed out of existence.

Project 2025, the right-wing blueprint for reshaping the US government, has a wide array of policies targeting climate change, none of them intended to stabilise a heating planet. Instead, the conservative wish list, published in 2023, calls for more oil and gas drilling. Donald Trump distanced himself from it during his election campaign, but, in power, his actions frequently echo the plan, including escalating the extraction of fossil fuels in Alaska and withdrawing from the Paris climate agreement. Project 2025's proposal for flood insurance is to end the federal programme and leave coverage to the private sector.

That proposal, if the Trump administration goes ahead with it, would bring about a sorting of the coast by wealth. As the increased costs of a heating world manifest, and as the state retreats, the most vulnerable are likely to leave first. These are people with the least secure livelihoods and the most transient status – households living in rented accommodation for whom it might be easier to relocate rather than stay to face the risk. The wealthiest will be able to protect themselves, at least for now, covering the costs of rebuilding either from private insurance or from their own resources. A section of society might find themselves in the middle, with just enough of their wealth tied up in property to make it difficult to retreat, but not enough to cover the costs of a catastrophe.

For generations, the automatic response to the threat of floods has been to build higher walls. This is true everywhere from the Dutch coast to the Mississippi Delta. Part of the resistance to that has come from the environmental movement, but human interference with nature is increasingly understood to have financial consequences too.

In 2018, when the US Army Corps of Engineers proposed a $6 billion plan to protect Miami from storm surges, it provoked a hostile reaction – and not just from those who preferred the natural protection of coral reefs and mangroves. Property developers were distressed at the prospect of a 6-metre-high sea wall cutting across the gleaming blue water of Biscayne Bay, a shallow estuary that attracts playful manatees, hordes of colourful fish and billionaire mansion-hunters.

The plan was rejected and the army engineers went back to their drawing boards to come up with a revised plan, this time focusing on elevating hundreds of homes and flood-proofing government buildings, including schools used as storm shelters. The military also recommended a pilot programme that would study how natural measures could provide protection against storm surges. But it warned this would be a challenge because of limited space; wetlands and dunes require large tracts of land and, in Miami, high-rises are built right up to the waterfront.

Most societies struggle to manage long-term risks. The temptation is always to find a short-term fix that pushes the problem out of view for a while.

In the UK, Flood Re has deferred a crisis, but it's easy to see how one could develop. Interfering with the market has encouraged people to settle in the riskiest areas, yet home-owners and landlords are not required by law to install flood defence measures. One in ten of Flood Re insurance claimants is a home with repeat flooding, but even for these households, building a more resilient home is left entirely to personal preference.

A coastal storm that overwhelms sea defences could inflict serious economic damage, leaving insurers facing huge costs and householders struggling to pay mortgages. The ripple effects of that will be felt through the system and the state is likely to be forced to act, rescuing insurers and banks as it did after the financial crisis.

In the US, the government may withdraw from providing insurance and even from building sturdier infrastructure. The likely consequence is that individuals and communities will be left to pick up the pieces as best as they can. The cost of disaster will be heaviest for the poor and for small towns that can least afford it – those who are also the least able to put pressure on the government. But if the poorest members of society are simply cut off, banks and private insurers are likely to be shielded from the consequences of a catastrophe.

Alongside retreating from the challenge, or ignoring it until it becomes a crisis and then letting the government pay the bill – which is the outcome that appears likely in the UK – there is a third possibility: the history of disaster insurance shows that societies can adapt to reduce risk.

The Somerset Levels is one of the lowest parts of England, a flat land, around a fifth of which is below tide level. Its history is coded in its name: Somerset is the land of the summer people because the floods that soak the land in winter yield lush grass to pasture animals in the balmy months of the year. On the Steart Peninsula, which sticks out like a hitchhiker's thumb from the Somerset coast into the Severn Estuary, flood defences have been deliberately breached and land given back to the water. 'The tide is able to come and go as it pleases,' as Alys Laver, the conservationist who manages the site, puts it.[14]

Turning farmers' fields back into a salt marsh went against the grain. In Parliament, the local MP accused the Environment Agency of spending 'far more money creating floods than averting them'.[15] The marsh is a haven for birds, especially in winter when flocks of dunlins whirl above it, bellies flashing white as they turn. The salt prevents the mud from freezing, allowing birds to hunt for worms and snails even in the coldest months.

But it is also an adaptive defence against a rising sea. When the defences were first breached in 2014, the entire salt marsh flooded at high tides, more than 100 times a year. Over the years, as silt has washed from the estuary into the marsh, the level of the land has risen – in places it rose more than a metre in the first four years. By 2019, the tidal waters covered the entire marsh fewer than 30 times a year.

In a 21st-century update of the London fire insurers who hired their own fire fighters, the British insurer Aviva is funding the restoration of a salt marsh on the Awre Peninsula on the River Severn in the Forest of Dean, modelled on the recreation of Steart marshes. The insurer is funding the project as a charitable donation to the wetlands charity WWT, rather than as part of a commercial relationship, but the experiment will help demonstrate whether natural flood management of this kind can play a part in reducing flood risk.

Successive British governments have set targets to build hundreds of thousands of new homes a year, a necessary step to keep pace with a growing population. But as houses are being built in vulnerable places and without preparing for the risk of flood damage, the risk of a disaster turning into a financial crisis grows. The rising cost of insurance is an early warning of the need for change.

After the flood at his home in north London, Chris Hook was glad to have saved some of the family's possessions. But he was stoical about what he lost, too. 'There was a moment of realisation that, if that hadn't been possible, would it have mattered that much?' Hook said. 'Not in the grand scheme of things.'

The first step is for people to understand the risks they face, but adaptation is likely to require deeper changes to the way we live. It isn't going to be possible to return to the lives of our ancestors, whose societies could be more mobile and more flexible in the face of disaster. Modern life involves living in megacities with complex and vulnerable infrastructure: power lines, water pumping stations, underground transit networks. People are just beginning to imagine what this new world might look like.

CHAPTER 11

THE RISING TIDE

By the sea, the island is wild and wind-ruffled, with a breeze blowing colour into walkers' faces as gulls complain over the grey water and industrial barges slide up the Thames. Perched on the seafront is a modernist cafe, a drum-shaped building with curving wings built to resemble the bridge of a cruise liner. In the way of maritime places, it's both a close-knit town and welcoming to strangers.

Slip behind those high walls and Canvey Island is suddenly closed off to the water, the North Sea and the Thames both invisible behind concrete. Smallgains Creek, where Mike Brown lived on a houseboat, was dammed after the 1953 flood and is now a playing field, a tongue of green surrounded by houses. The high street is familiar English suburbia; charity shops, takeaways, tanning parlours.

Canvey is far less vulnerable to a storm surge than it was in 1953, ringed by a 23-kilometre stretch of flood defences. But the sea is not the only threat it faces. Stand on the landward side of the high encircling wall and the town lies in the bowl of a saucer, which could fill up in an instant if there is a torrential downpour.

This is exactly what happened in the summer of 2014, when 10 centimetres of rain fell on the town in four hours and hundreds of houses were flooded. 'It looked like the 1953 floods all over again,' according to Dave Blackwell, an islander who experienced both inundations.[1] That day in 2014, it stopped raining and the sun came out. 'Kids were out swimming in it,' Blackwell said. The rain held off and houses were pumped dry.

The island is a fortress, but one that is vulnerable to a few too many drops of water. The Environment Agency spent £75 million in 2025 strengthening sea defences that directly protected 6,000 homes on the island.[2] Put crudely, this is a cost of more than £12,000 for each household. The town is already 2 metres below the high-tide level of the Thames estuary; a rising sea will require further strengthening and raising of the fortifications.

The heroic efforts of Reg Stevens and others restored Canvey to life after the 1953 storm, but, in its war with the elements, the Essex island reached a fragile stalemate. In a way, that makes it hard to tell whether Blackwell is joking when he says, 'You could give all of us a few hundred thousand and move us off.'

On the other side of the North Sea, the Amsterdam neighbourhood of Schoonschip is an experiment with a new way of living with water.[3] On a sunny day, when children race around on bikes and toss frisbees for dogs, there's little sign that this is an unusual neighbourhood. Schoonschip is a community of floating houses: cube-shaped blocks with tall windows, wooden cladding and solar panels on their roofs, afloat in one of the city's canals and linked to the shore by wooden jetties.

Inside, each of the cubes contains two households, in separate houses that are arranged across multiple floors, with the lowest level below the water line. Each one is different, but they are spacious and elegantly decorated, with polished concrete floors and wooden staircases. Folding doors open onto sun-filled terraces and living rooms drop down to outside decks where residents dangle their feet at the water's edge. With mooring posts pinning them to the canal bed, the cubes don't move more than a fraction to either side. In a storm they bob a little, and dangling ceiling lamps inside the houses sway, but everything else stays in place. Glasses don't slide off tables.

Floating houses like this still require protection from damage by floodwater. A powerful storm surge would be just as destructive here as it would be on land, but Schoonschip bends in the face of more modest disruptions: the houses can rise around 20 centimetres or fall the same amount as the level of water in the canal rises or ebbs. More violent disruption would risk damaging the connections that link Schoonschip to the power grid and bring in clean water in pipes that run under the wooden decks around the buildings. It's a tiny community, just 46 households, but Schoonschip offers a template for living in a way that embraces rather than resists water.

For centuries, people have adapted to life in swamps and on river deltas with floating homes, or houses raised on stilts. The floating houses of Schoonschip combine humanity's past with a vision for the future. New technologies have transformed cities before. For instance, the invention of modern safety elevators and the use of steel frames allowed the creation of skyscrapers, letting cities like Chicago expand upwards.

A floating house linked to the power grid and water systems creates a way to redesign cities in weeks or months rather than years. Floating blocks can be built in factories and towed into place, creating new neighbourhoods where they are needed or removing them where populations are shrinking. Student accommodation or old people's homes can be added as the need arises.

Legal questions might be a bigger challenge to this vision than practical ones. A few years ago, a Dutch architect designed a floating mansion with steel piles that drop down to moor on the seafloor. Called the Arkup, it's a white rectangle that extends over two storeys, a villa on the water just off the Miami coast. The question for Miami's tax office – and the courts – was whether this was a house, and therefore eligible for property taxes, or a boat. A tax bill for nearly $120,000 turned on the answer. A judge ruled that it was a vessel and therefore not subject to property tax, but the victory could prove a setback for the idea of building on water: a house that can sail out of the taxman's reach is unlikely to win favour from governments.

Floating neighbourhoods are only likely to work when the political and economic rationale coincide. As in the 17th century, when Dirck van Os and fellow investors drained the Beemster polder, the creation of a new landscape will be driven by the profit motive of entrepreneurs, alongside a country's quest for space; gulf states like Saudi Arabia have both the maritime technology and the need to house a growing population.

The legacy of the 1953 North Sea storm was an emphasis on using technology to resist nature. With the shadow of a titanic war behind them, the British and Dutch pursued an engineered response to the challenge of living by an unpredictable sea. Both the Thames Barrier and the Delta Works were breathtaking in their stature and the skill they took to build – gigantic piles of concrete and steel that are precisely tailored to the shape of the estuaries they defend.

The barrier at Woolwich is just part of a network of defences protecting the UK's capital, hundreds of kilometres of walls and embankments along the river and more flood barriers with moveable gates – the Thames Barrier's little siblings – in tidal rivers and inlets that flow into the Thames.

It is formidable, but like any piece of machinery, the more it is operated, the more maintenance it requires. As sea levels rise, increasing the height of daily tides and the potential size of storm surges, it will need to be closed to defend the capital more often. Fifty closures a year is the barrier's safe limit. Above this, there is insufficient time for maintenance. The barrier is likely to hit this level by 2035.

The Maeslant barrier defending Rotterdam was designed to last a century. When it was completed in 1997, the surge defence was capable of handling half a metre of sea level rise. But the accelerating rise of global mean sea level means its likely lifespan is shorter.

There are ways to extend the life of a barrier. Many of the Thames Barrier's current closures are to protect from flooding forming from upstream, not from the sea but rain swelling the Thames along its vast catchment and flowing down to west London. If defences upstream along the river are strengthened, the barrier will not need to be closed as often.

But in time, even defensive works as impressive as the Thames Barrier will no longer be sufficient. A few years ago, a Dutch oceanographer proposed an extraordinary response to sea level rise: the enclosure of the North Sea with two giant dams, one stretching nearly 500 kilometres between Scotland and Norway, and the other traversing 160 kilometres of sea from southern England to France.

Sjoerd Groeskamp, working with a colleague at a German university, suggested it would protect 25 million people in 15 countries around the North Sea and Baltic coasts, including the Netherlands and southern England. This mammoth successor to the Delta Works was intended not just as a solution, Groeskamp said, but as a warning of the scale of the threat. But turning the North Sea into a lake would cost billions of euros. Both the scale of the finance required and the complexity of its politics – it would require close cooperation between every country affected, from Britain to Russia – mean it is unlikely to happen.

Storm surge barriers and sea walls provide a powerful sense of safety, but they are an all-or-nothing defence – a single failure in the face of a colossal storm would be devastating. The growing cost of building defences such as these in the face of rising seas, alongside growing concern for the environment, has forced a reassessment. By the 1990s, planners in the UK had accepted that coasts were a living system, and that shoring up a cliff in one location would prevent sediment being carried along the shore and lead to the erosion of beaches hundreds of kilometres away.

Over the span of a single human lifetime, a landscape might look static, but over the long timescales that scientists who study the earth work to, it's not a still image – it is cinema. Villages and sometimes entire cities disappear beneath a flood. Humans reshape nature and then the sea and river reshape us.

To say that the coast is alive is a metaphor, but it is also a literal truth. Land and sea are in motion and the relationship between the two is altered by the plants and animals within it, as well as by us. Dunes are held together by marram grass. There is no fenland without slowly decomposing mosses and other wetland plants. Oyster reefs are built out of living creatures growing shell on top of shell. Plants and creatures create landscapes that absorb water and buffer the energy of waves.

Sea walls and river dikes aren't just physical barriers, but mental dividing lines too. Creating 'room for the river' in the Netherlands improved people's quality of life by connecting them to the water, creating swimming spaces and new walking routes in the middle of cities. Walls impose an artificial order on the coast, making them static, unbending lines. A marsh is, literally, a more fluid approach to living with the sea's changing temper. Maarten Kleinhans, professor of geosciences at Utrecht University, suggests that instead of thinking of coasts and rivers as lines or shipping routes, we could see that a continent is a river. Rivers don't just flow through the land. Instead, the water gathers on nearly every surface before trickling down into vast rivers like the Rhine or the Thames.

Juxtapose the cost of engineering and the beauty of a salt marsh, and the answer seems obvious: a reversion to nature along every inhabited coast. But it isn't that simple. For natural flood defences to work, nature has to be given room – just as the Dutch did with their rivers – but cities that are densely packed and commercially valuable have little space to yield.

Instead, we have to balance the restoration of nature with technological solutions. In Rotterdam, the Maeslant barrier defends the port city from storm surges, while, behind it, the Dutch have increasingly adapted their architecture to accommodate rather than resist the water.

At Benthemplein, a square near the city centre, a sports pitch is inset into the middle of a plaza with steps leading up from its sides, like a Roman amphitheatre. When it's dry, skaters and basketball players dart around the pitch, and when there's heavy rain, it fills up and turns into a pond.

There's grass rather than asphalt between the tracks of Rotterdam's trams, so rainwater is absorbed rather than spilling onto the streets. These adaptations are minor, and the city is still a maze of concrete and steel, but they indicate a change of mindset.

Early this century, the Netherlands established a Delta Commissioner, the official's title an echo of van Veen's Delta Committee, who is responsible both for defence against floods and preparation for water scarcity. It is an unusual appointment, because governments usually prepare for a crisis only after it has happened.

Cornelis Verdaas, who was appointed to the role in 2023, has to tell farmers that, in a few decades, tulips, the vibrant flowers that are both a national emblem and lucrative export, will no longer exist. This is because of a longer dry season and a higher sea, which means salt water intruding into the soil of agricultural regions in the west of the country. 'We have 20 or 30 years to adapt,' Verdaas said.[4] 'This is not a nice message, but an honest one.' He is convinced that the Dutch have the resources and expertise to cope with sea level rise but, like many officials around the world charged with flood defence, wonders about the psychological impact of severe floods. 'Professionally, financially, we can manage,' Verdaas said. 'But we cannot manage how people will respond to an uncertain future.'

Dutch officials say they will never give up the 'Randstad' – the four cities of Amsterdam, The Hague, Rotterdam and Utrecht on the low-lying western edge of the country that have merged into a single megacity – because that is the heart of their manufacturing, finance and trade, where they earn the money to pay for protection. But new housing will need to be built on higher ground and existing homes will need to adapt to floods, accepting that water may sometimes enter and need to be cleaned out.

Decisions taken deep in the past have shaped our present vulnerability, from the ancestors of the Dutch settling in a river delta to the French colonists establishing New Orleans on the banks of the Mississippi. The challenge now is to shape a future that allows our descendants and the world around them to thrive.

Long ago, humanity co-existed with nature. This was partly out of necessity, as ancient societies lacked the technology and complex finance needed to transform their landscapes. But there was a philosophical strand to this too. Nature was not just ours to harvest; a wise steward understood that it needed protection. The Romans and other ancient cultures trained grapevines on trees, and released pigs to forage in woodlands, creating an artificial version of the mosaic of a natural forest. River floods could be destructive, but in their wake they brought fertile black silt and lush grass.

Our relationship with seas and rivers was transformed by the diking and draining that the Dutch pioneered and perfected, and financed by the capitalism that was invented alongside. The Industrial Revolution distanced humanity from the natural world, fostering the rise of giant cities. Egypt may once have been Kemet, the black land, named for its fertile silt, but damming the Nile trapped the river's life-giving sediment and left farmers reliant on artificial fertiliser. Britain was a much damper place in the past, with a thick coating of temperate rainforest along its western flank, while wetlands have vanished everywhere from Yorkshire to Devon, a lost landscape that once squawked, grunted and screeched with the noises of beaver, boar, heron and marsh harrier.

It's obvious that cities are artificial constructions, but the same is true of the countryside, where wetlands have been erased, rivers that were once shallow and wide have been given steep edges, and 'groynes' – those concrete structures projecting out to sea – limit the movement of sand and shingle along a coastline. The taming of nature hasn't just protected cities from floods; it has created fields of wheat instead of boggy fenland, rivers that are high and swift where houses can be built right up to the edge, picture-postcard sandy beaches that tourists love.

Trees have been felled, hedgerows ripped out, marshy land drained and clay pipes laid beneath the surface to speed the flow of water away from farmland. Working fields with heavy farm machinery when the soil is damp leaves it more compact. This has consequences in heavy downpours, when trees and hedges with roots that once held water are gone, the soil is less porous and rain cascades down bare valleys to swell the streams that run below, sweeping down on towns and villages.

The 1960s brought a greater appreciation of the environment, a renewed recognition that our species is part of nature, but the pattern of flood defence had been set. After every crisis, there was a scramble to build higher defences.

There's a techno-optimist view of this, one that says humanity can solve every problem we have created simply by applying more technological ingenuity: building higher sea walls as glaciers melt. But as the climate alters and sea levels rise, our ability to adapt will be tested to the limit.

'We can't keep investing more and more to protect everybody,' an Environment Agency official said.[5] 'Most people don't know they are at flood risk. The perception is that people who live on Mill Street or Pond Lane are never going to be flooded.' Officials and scientists who study flood defence describe this as a shift to resilience rather than resistance –

accepting that floodwaters may wash through a home, but installing plug sockets higher up the wall and removing carpets from ground floors.

This will mean making choices. The East Riding of Yorkshire begins at Spurn Point, the headland where the 1953 storm made landfall, and runs 85 kilometres north along cliffs of rapidly eroding boulder clay. The north Norfolk coast runs for more than 100 kilometres east from Hunstanton, where US airmen led the rescue in 1953, to Lowestoft.

These are the fringes of England where choices about resettlement and retreat are likely to come soonest. Dealing with floods has never just been an engineering question. It has always, just as much, been a question of what society values and why. As Hermann Bondi wrote, the risk to London was the potential of a 'knock-out blow to the nerve centre of the country', but the UK can survive without Canvey Island. That choice will be outlined in stark terms as coastlines are increasingly menaced by the sea.

In 2013, the Environment Agency allowed a shingle bank that protected the coastline to erode into the sea at Medmerry, 11 kilometres south of Chichester in West Sussex. Giving up the sea defence allowed the tide to flood acres of farmland and turn it back into saltmarsh. A new sea wall, a 6-kilometre-wide clay embankment, was built a short distance inland. Officials described the flooding at Medmerry as a 'managed realignment', but in plainer language, it was a retreat. It is the largest area of coastline that has been given up to the sea anywhere in Europe.

Within the coming decades, these choices will be imposed on coastlines where cliffs are crumbling, sea levels are rising and there are fewer people, properties and businesses to protect. Britain's first climate refugees are not likely to come from its capital. A drift from the coast may already have begun; the last census indicated that north Norfolk's population was growing much more slowly than the rest of the country, while the number of working-age people and children was declining.

In January 2020, during Donald Trump's first presidency, he described a plan to spend an estimated $119 billion on a surge barrier to protect New York City as 'costly' and 'foolish'. He added that, when needed, the barrier 'probably won't work anyway'. Six weeks after the president's tweet, the Army Corps of Engineers announced that further work on the project was 'indefinitely postponed'.

This barrier would have reached from Queens to a large spit of sand extending from the coast of New Jersey, spanning 10 kilometres across New York's outer harbour. Its design resembled the Eastern Scheldt storm surge barrier in the Netherlands, a succession of retractable gates that would swing shut if a storm surge approached the city and threatened a repetition of Hurricane Sandy's devastation. Alongside this, there would be more than 40 kilometres of floodwalls, levees and dunes along the shoreline.

Well-maintained surge barriers have repeatedly proved able to withstand fierce storms, but the president's anxiety about costs – even to defend the financial centre of the wealthiest country on the planet – indicates that the era of giant flood engineering projects may be coming to an end.

After Sandy struck the city in 2012, New York's mayors authorised sweeping changes to its building codes: new buildings in flood zones had to be fitted with backwater valves that prevented sewage from flowing back into homes; in new healthcare facilities, the rooms where patients slept had to be located above expected flood levels; new commercial buildings have to be be elevated at least 30 centimetres above the anticipated level of future floods. On the eastern shore of Staten Island, hurricane-damaged homes were bought out by the state government and returned to wetland.

Even without the protection of a surge barrier, these changes are creating a more resilient city, but New York's exposure remains vast. According to one study, up to 82,000 homes in the city and the suburbs to its east could be lost due to flooding by 2040.[6]

Across maritime Asia, cities on coasts and in river deltas face grave danger. The Indonesian capital of Jakarta is gradually disappearing below sea level. When it rains, the rubbish-clogged drains are quickly overwhelmed and the giant city's streets and houses are regularly flooded, in rich and poor quarters alike. The water is frequently deadly: people drown or receive electric shocks from cables dislodged by floods. Jakarta's problems are amplified by climate change, but they are in large part man-made.

Water flowing down into the city now encounters extensive paved surfaces and rivers that have been diverted from their natural flow into canals, while people seeking drinking water have dug wells that drain the underground aquifers beneath the city, lowering the whole metropolis. Ironically, it is the search for water to drink that has made the city most vulnerable to being flooded. Removing concrete surfaces, allowing its rivers to bend and retain water, and – crucially – putting a stop to the extraction of groundwater could revive the city's fortunes. Instead, Indonesia's government plans the construction of a new capital more than 1,200km away on the island of Borneo.

Returning to a simpler past may be impossible, but our species could use its technological power to restore rather than destroy nature. A deep-rooted belief that nature can be managed now has to be balanced against what science tells us about the increasing challenge from rising seas.

The restoration of nature and engineering can intertwine: instead of sea defences that present unyielding concrete cliff faces, 'living sea walls' have been built along the Dorset coast with concrete basins projecting from their seaward sides that mimic rock pools, creating a habitat for marine life. When wetlands are destroyed, they are rarely recorded, but remote sensing from satellites and planes can be used to map the topography of lost marshes and help draw up plans to restore them. Noah's ark is not just a story of humanity rising above the flood. It's also a story of mankind using technological ingenuity to protect the rest of creation.

Moderating the impacts of climate change will be crucial to securing a liveable future. Even if sea levels continue to rise, curbing greenhouse gas emissions will avert the worst extremes of storms and rainfall. But there's no question that the outlook for cooperation on climate change is now bleaker than ever. The outbreak of war in Ukraine and Gaza has created suspicion where humanity needed solidarity. China and India are ignoring the West to do energy deals with Russia, while the US is focused on extracting and exporting fossil fuels.

If there is hope, it lies in the spread of renewable technology, including the adoption of electric cars, which means we appear to be close to the peak of oil consumption for energy. There is hope, too, in our spiritual connection to the natural world. The last few centuries have showcased our appetite for destruction, but we have an enduring need to find healing and pleasure in greenery, and the hopping, fluttering life around us.

Penny Dack was five when the 1953 floods hit Lowestoft. She remembers her father, George, a policeman, coming home exhausted in his long, black, police-issue raincoat and rubber boots.[7] He had spent the night evacuating the families of Beach Village, a tiny fishing community at the foot of the cliffs that run along the town's seafront. It's a place that no longer exists; the flood dealt a final blow to a community the authorities deemed a slum.

There were fish curers, smokehouses and fishing families living in two-up, two-down tenements. Scottish 'fisher girls' would be a regular sight, women who followed the migration of herring down the coastline and found work gutting and packing the fish caught by their husbands. Their hands were always busy: mending nets, knitting jumpers. Beach Village was a fragile sliver of life by the sea, not beautiful exactly, but filled with industry and excitement: it smelled of smoked fish and the tang of sea air.

As a teenager, Penny would come down to the cliffs when she felt anxious, gazing at the moon's reflection in the water. The sound of the sea would wash her worries away and she would stay until her dad came down to find her in the early hours of the morning. No one died on land in Lowestoft during the 1953 floods, though the houses of the Beach Village were flooded and a trawler was lost with all its crew in the North Sea. The threat of the flood is ever-present here, but the sea's rhythms offer solace too.

Penny now lives in a market town around 16 kilometres inland from Lowestoft, in a house that's on higher ground and far from the nearest river. Like many people who can recall a catastrophic flood and its aftermath, she is both conscious of the risk and a little less attached to material things. The sea brought healing when she was a girl and now happiness is a walk with her spaniel, Issy. 'Seeing what's around me, smelling what's around me,' she said. 'When I see litter, it saddens me. When I find an area that's been greened, it boosts me enormously.'

Lowestoft has water at its heart. A saltwater lake splits the town in two and flows out to the North Sea, making it impossible to live here and not think about the sea. Bridges link the two sides of the port. The newest is the Gull's Wing Bridge, a kind of construction known as a 'bascule bridge', with wings that lift up to let shipping pass underneath, like London's Tower Bridge. Fishing no longer makes up much of the town's income after herring was over-fished and stocks of the species plummeted in the 1960s, taking the livelihoods of the fisher girls and their menfolk with it. But the connection to the sea remains, as the port services the offshore wind industry.

When Storm Xaver tested England's sea defences in 2013, Lowestoft's harbour overflowed and more than 150 homes and around 200 businesses were flooded. Government engineers strengthened flood walls lining the roads around the harbour, but plans to build a tidal surge barrier across the harbour's mouth – a moveable defence 40 metres high which would have been the second biggest in the country after the Thames Barrier – had to be abandoned for lack of money.

Instead of a flood defence system, one resident said, Lowestoft had been left with 'what looks a lot like a funnel'. A surge sweeping in from the North Sea would not be stopped at the harbour entrance, but would thrust deep into the centre of the town.

Penny's son-in-law, running a company that manufactures steel products, was one of those whose businesses were flooded by Xaver. Alongside the cost of repairing the flood damage, he found it harder to get flood insurance afterwards. Without a guarantee of flood defence, Penny fears a vicious circle in which it becomes more costly to get insurance, more difficult to get mortgages and property prices spiral down. 'Lowestoft is on the front line,' she said. 'We want to regenerate and rebuild. But without a tidal barrage, that's not going to be a good idea in the low-lying part of Lowestoft.'

There's a hard-headed case for spending millions to protect Lowestoft – the argument that the town's revival lies in becoming a centre for the offshore wind industry, part of a greener future. And there's a romantic one too, which says that this town and its long kinship with the sea is part of England's heritage, that future generations of girls should sit on cliffs and stare at moonlight in rippling water.

CHAPTER 12

FUTURE WEATHER

Until recently, maps showed very little of Antarctica, a blank at the edge of the world. Explorers sighted its mainland in the early 19th century and the allure of the last entirely unexplored and untouched wilderness drew bold spirits, seeking the glory of conquering the blank by planting a flag at the southern edge of the world.

For ages, the Antarctic has slumbered beneath its deep blanket of ice. Scott, Amundsen and Shackleton's race to the ultimate south pinned it in the public imagination as a place of heroism, a last flowering of the romantic age of exploration. It is most famous for its stoical, doomed voyagers. Aware that his frost-bitten feet were not healing and not wanting to be a burden, Captain Lawrence Oates told his companions 'I am just going outside and may be some time,' before walking out into the snow.

The vast continent of snow, ice and rock is better known to humanity than ever now, with an age of heroes giving way to cool scrutiny from above by satellites and from the sea by aquatic drones. Antarctica's future has been charted with computer models and its substance sampled with drills into the deep ice. And while many questions remain unsettled, what we know is that the giant is stirring – and what that means for us is that the blank space at the bottom of the map is preparing to reshape Earth.

Most of the planet's fresh water does not course through streams and rivers or gather in lakes. It is bound up as snow and ice. Earth's frozen parts hold 70 per cent of all fresh-water, sometimes releasing it to flow downriver to the sea, as Pakistan's mountains do in summer.[1] The cryosphere, as Earth scientists call it, is in retreat. Glaciers and ice sheets are melting, the extent of spring snow in the northern hemisphere and Arctic sea ice is reducing, and temperatures are rising in permafrost. And the single largest part of the cryosphere, holding most of the planet's freshwater, is in the great expanse of ice above the southern continent.

Like the height of a mountain, sea level is a mapmaker's fiction – and one that arose with the modern world. New York is said to be 10 metres above sea level, but descriptions like this naturally depend on where exactly sea level is measured from. In Britain, sea level references are given with three initials after them: ODN, which stands for Ordnance Datum Newlyn, after the Cornish town where sea level was measured using a tide gauge. ODN has been Britain's reference point for sea level since 1921, barely a hundred years.

More ancient societies, faced with an entity as changeable as water, measured the height of the sea's fury with flood markers and the depths of the rivers' retreat with famine markers. In 1918, a 'hunger stone' was revealed on the bed of a central European river, with the words, in Old German: 'When ye see me, ye will weep.'[2] The extremes of water levels, which determined whether people prospered or suffered, were what mattered.

But for our time, when we often find it easier to think about maps and lines on maps rather than what they represent, sea level is a convenient marker. Humanity's transformation of the planetary climate is raising the seas, because more heat melts the ice on Greenland and Antarctica, which then flows into the oceans as water – and also because water expands as it gets warmer.[3]

The rise in sea level is not going to be exactly the same everywhere, as that depends on local factors, such as the strength of wind and ocean currents, which affect how and where heat is stored in the ocean. But, as a rough guide to what's happened since the dawn of the industrial age, scientists say the rate of sea level rise is accelerating from 1.4 millimetres a year through most of the 20th century to 3.6 millimetres a year between 2006 and 2015. The rise in sea levels means that storm surges driven by a future Katrina will push further inland, risking greater destruction, at a time when increasing numbers of people are living by the coast.

If time is measured in human lifespans, sea level rise is a leisurely affair. Globally, the seas are expected to rise by around a metre more by the end of this century, with a chance of a greater increase without effective efforts to reduce emissions.[4] From the perspective of Tuvalu, a Pacific island nation that lies around 4.6 metres above sea level at its highest point, this is far from leisurely, of course. But it is a timespan which allows nations at higher latitudes to make critical decisions about the future.

And the decades until the year 2100 are not such a long time when measured in the span of a city's existence; London was founded around two millennia ago and New York has been around for four centuries. The questions that scientists are now trying to settle in Antarctica will tell us how long we have to prepare – and exactly how much of a challenge humanity faces.

In 1978, the British scientist John Mercer warned of a disastrous rise in sea levels from the melting of the West Antarctic, fearing the drowning of Florida and the Netherlands, a prophecy dismissed by one scientific journal as the stuff of a low-budget movie.

Mercer's prediction was alarming, but the study of Antarctic ice is a new field that was still in its infancy in the 1970s. Satellites usually orbit over the Earth's equator, making the poles a blind spot and the first satellite to provide comprehensive coverage of Antarctica was NASA's Nimbus 7, which was only launched in 1978, months after Mercer's paper was published.

While satellites have vastly improved human understanding of Antarctica, they have limitations; a satellite can only detect the surface of the ocean or the ice rather than what is happening beneath it. And a satellite will measure the thickness of floating sea ice by bouncing radar off its surface and comparing this with the open water, but its radar measurements can be confounded by compacted snow lying on top of the ice.

Scientific expeditions which camp out on the ice during the Antarctic summer provide an opportunity to gather data from deep beneath the surface. Observation of the Antarctic began with some of the earliest explorations. Scientists on Scott's Terra Nova expedition of 1911–12 kept detailed accounts of pressure, temperature and the wind's direction and force, along with their impressions, which were sometimes less scientific. In the register, one of the men wrote, as they endured their first Antarctic winter, that it was 'blowing like hell'.

From the 1960s, researchers began extracting cylinders of ice from Antarctica's glaciers. Glaciers take many years to form as older snow is buried by new snowfall and is slowly compressed into a mass of ice. Pockets of air trapped in the ice reveal the composition of the atmosphere deep in the past, stretching back hundreds of thousands of years.

In this deep freeze, there is an archive of the planet's climate that predates the observations being gathered from the atmosphere at weather stations like Mauna Loa. The evidence of the ice cores shows that the concentration of carbon dioxide in the atmosphere had been steady for the last 1,000 years, before levels of the gas began an abrupt and steep climb in the 19th century.[5]

Alongside satellite sensors and field exploration, the third way to comprehend Antarctica is through computer modelling. This is the only way to forecast possible futures.

The southern continent is divided in two by the Transantarctic Mountains, a rocky spine stretching more than 3,000 kilometres. The eastern portion of Antarctica, due south of Australia and the Indian Ocean, is much bigger than the west and much of its ice rests on bedrock high above sea level, which makes it much harder to melt. It is still possible for the eastern Antarctic to be melted by warmth in the atmosphere, but this is a much slower process than being heated by the ocean.

That, in simple terms, is because water is much more efficient at transferring heat than air. It is good news for humanity. If that vast quantity of ice east of the Transantarctic Mountains turned into water, it would raise sea levels more than 50 metres.

The West Antarctic is a different story. It is, by its nature, much more precarious. Much of its ice rests on bedrock that is below sea level, a significant part of which is on 'retrograde' bedrock that slopes downwards towards the centre of the continent, like a shallow bowl. A glacier is a river of ice, formed where snow accumulates in the cold interior of Antarctica which then gets compressed down into ice and flows out to the sea at a languid pace. At the point where the glacier reaches the edge of the continent, it feeds an ice shelf, a tongue of ice hanging out over the sea.

When the sea melts a glacier that's flowing along a retrograde bed, the ice retreats back down the slope, back down towards the centre of Antarctica. As it does so, more of the cliff-face of ice at the front of the glacier is submerged in the water and exposed to its warmth. That creates a feedback loop. There is further melting. Even if the speed at which the glacier is pushed out from the interior of Antarctica stays the same, the total volume of ice lost keeps growing.

Two giant glaciers in West Antarctica are currently retreating on retrograde bedrock. The Pine Island glacier reaches the Amundsen Sea near the base of the Antarctic peninsula, the thumb of land extending due north towards South America. This glacier is one of the fastest-retreating on the entire continent and it drains an area of more than 162,000 square kilometres.

Its neighbour, the Thwaites Glacier, is even bigger, a behemoth that is around 120 kilometres wide and drains an area of about 192,000 square kilometres, about the same land area as the UK. Thwaites extends into the interior of the continent; the loss of these two glaciers would not just raise sea level by itself, but risks destabilising the entire West Antarctic ice sheet. Between the 1990s and the 2010s, the quantity of ice flowing into the sea from Thwaites and neighbouring glaciers more than doubled.[6]

Deep beneath the ice at the edges of Antarctica is an answer to one of the biggest riddles facing humanity. It can't be seen by satellites' instruments and is too far from the front of the ice to be reached by underwater robots. The place to begin looking for answers is by drilling down from the surface of the ice. The fact that the ocean drives the melting of ice is understood. Less clear is why the rate of melting can vary so much within distances of a few kilometres. And where exactly Antarctica is melting is what matters.

The grounding line is the point where the ice loses contact with the bedrock and first begins to float. This can exert a force back upstream, perhaps because the ice is in contact with the seabed, and that force works like the buttresses that support a cathedral. If the buttressing force reduces, because ice is melted away at the right point by the ocean, the flow of ice off the land can speed up. Peter Davis, an oceanographer at the British Antarctic Survey, puts it like this: 'In some parts of a house, you can knock a wall down and nothing happens. In other parts, you knock down a wall and the whole thing comes crumbling down.'[7]

Studying Antarctica in person is an activity suited to the resourceful, both physically and psychologically. The continent is vast and hostile, and offers little to soothe the human eye. In sunshine, there are undulations in the snow, like ripples of sand in the desert, but on an overcast day, it can be a featureless expanse of white. Opportunities to stay in touch with family members back home are slender; the focus is on the scientific mission. That, and staying alive. The scientists must remain fit and healthy for months, preparing their own food and sleeping in pyramid-shaped tents buffeted by fierce winds that whip up particles of ice.

Davis was one of a team of researchers who drilled through the ice of Thwaites in the Antarctic summer of 2019–2020. The glacier contains enough ice to raise the world's sea levels by 65 centimetres if it melts and its grounding line has retreated inland 14 kilometres since the late 1990s. The results of the drilling revealed a more subtle change than expected. Although the ice was melting, the rate of melting on flat parts of the ice shelf was slower than predicted by computer modelling. But, in terraces and cracks on the underside of the ice, rapid melting was happening.

How the ocean and the ice interact varies by location. The challenge is to take observations made at small scale – measurements obtained from sending probes through holes drilled in the ice – and use them to develop large-scale models that can guide governments on planning for the end of the century. When Mercer raised the alarm, Antarctica was still largely unmeasured. The combination of satellites and researchers in the field has provided the ability to quantify how much the ice has retreated.

Scientists are open about the uncertainty around Antarctica's trajectory, but researchers who study the continent's ice are converging on the idea that large parts of the West Antarctic ice shelf will collapse over the coming centuries, with disagreement focusing on how rapid this will be. Not if, but when.

Ice sheets move to a slow time scale. Kaitlin Naughten, an ocean modeller and one of Davis's colleagues at the British Antarctic Survey, said: 'They have a huge amount of inertia, more so than any other element of the planetary system. Very large and very viscous fluids respond slowly, then build up their own inertia, and take a long time to stop.'[8] The giant takes a long time to wake to full alertness.

This has long-lasting consequences. Even if humanity succeeds in curbing emissions and global temperatures stabilise, the ice will continue to melt and sea level will continue to rise for centuries. The whole of the West Antarctic ice sheet contains sufficient ice to raise global sea levels by 5 metres.

Climate change can be imagined as a titanic act of engineering, a transformation of the planet from one state to another. Turning the water back into ice, sending the giant back into his freezing state, will be one of the slowest parts to unwind. Even if attempts to mechanically remove carbon dioxide from the atmosphere are successful, the process set in train now will take a long time to unwind.

Modelling by Naughten and colleagues has found that rapid warming of the ocean and subsequent sea-level rise appears certain this century. Scenarios in which mankind reins in emissions made no significant difference. But the effects can be amplified. An extreme scenario in which humanity emits much more carbon into the atmosphere would accelerate the collapse.

There is a further twist. Ice sheets exert a gravitational pull on the sea around them, which weakens as they melt. This means sea levels near melting ice sheets drop, rising further away. The waters around Australia and New Zealand will be influenced by the melting of Greenland's ice sheets. Antarctica, the giant that sleeps at the bottom of traditional maps, is perfectly placed to raise sea levels in northern Europe.

The scale of the threat is beginning to make extreme solutions look more plausible.[9] John Moore, a glaciologist at Beijing Normal University, and Michael Wolovick, a geoscientist, have proposed constructing giant ridges of rock extending from the seafloor which could pin the West Antarctic ice shelves in place and buttress the glaciers behind them. Engineering firms have taken part in discussions about constructing 'curtains' anchored to the sea floor and made out of seawater-resistant materials that would block the flow of warm water that is melting the ice.

These ideas would face immense practical challenges. It is difficult enough to sustain a scientific mission on Antarctica, let alone engage in a massive construction effort in the iceberg-strewn seas off its coast. Interfering with the continent's environment, so far barely touched by humans, would also inflict significant disruption on its marine life.

The logistical challenges are ones that humanity may be able to solve. The biggest reason to file proposals like this under fantasy is because they would require cooperation between the world's powers. The US, Russia and China all have voting rights under the treaty system that governs the Antarctic. And if those nations increasingly see each other primarily as rivals, it's hard to imagine what it would take to make them put their differences aside.

The Thwaites glacier is sometimes described as the 'doomsday glacier', but this may be a misleading way to think about it as we tend to think of the apocalypse as a sudden and dramatic event, rather than a slow shift to a new state of affairs. Measured in the pace of human lifetimes, a rising sea will mean successive generations grappling with difficult choices about rebuilding or relocating further inland as coastal settlements face an increased threat of flooding.

One reason it is difficult for humanity to adjust to this new reality is that for nearly all of recorded history – the past 6,000 years – sea levels have been stable. Around 15,000 years ago, when ice sheets covered Britain and prehistoric humans stalked mammoths, the seas were around 130 metres lower. As the climate warmed, driven by changes in the planet's tilt and shifts in its orbit, the glaciers melted and sea levels rose.

With the invention of agriculture, as populations boomed and cities grew bigger, the coast was a safe place to live. A flood would sometimes cause devastation, but the level of threat was not rising year by year. And as populations grew over the last two centuries, the flat land and trading connections of coasts and river estuaries made them natural places to build. More than two-thirds of the world's mega-cities, such as Mumbai, Lagos or Tokyo that have upwards of 8 million inhabitants, are on the coast.

The stability that our ancestors enjoyed has now gone. Even after the world reaches net zero, and the other effects of climate change stabilise, the sea level will continue to rise for centuries.

A large part of the future will be written by rising seas, but there is another way in which the water molecule will shape human destiny – and that is from the sky. Over the last few decades, alongside our growing knowledge of the world's snow and ice, we have deepened our understanding of the atmosphere – and humanity's impact on it.

Guy Callendar in the 1930s and Gilbert Plass in the 1950s had both investigated the effects on the atmosphere of carbon dioxide, with Plass calculating that the Earth's temperature would increase by 3.6° C if the amount of carbon dioxide in the atmosphere doubled. But their work had not taken into account the effect of heat being transferred from the Earth's surface as hot air rises; this effect is visible to the human eye on a hot day as a shimmer.

In the mid-1960s, a Japanese physicist Suki Manabe and his colleague Richard Wetherald, both working at a government laboratory in Princeton, New Jersey, developed the first computer model of the climate. This was a time when weather forecasting was being revolutionised by the use of computing.

Through most of history, it had been a matter of judgement and experience more than science, with forecasters using their knowledge of patterns from the past to assess the future. During the First World War, working amid the carnage and mud of northern France, a British mathematician called Lewis Fry Richardson had devised the first numerical method for predicting the weather.[10]

He worked by hand and it took him six weeks to come up with a forecast for six hours of weather in a single location in Europe. His simulation of the weather turned out to be inaccurate, but Richardson's method laid a path for others to follow.

As well as the sun's radiation, Manabe and Wetherald's model included the transfer of heat from the Earth's surface to the atmosphere and the evaporation of water.[11] Their 1967 paper found that a doubling of CO_2 in the atmosphere would result in an increase of temperature at the Earth's surface by $2.36°$ C. This estimate of how sensitive the planet is to rising carbon dioxide levels is in line with the current science.

The researchers also predicted that, as the ground and lower atmosphere warmed with increased CO_2, the temperature would fall in the stratosphere – the dry and cold upper atmosphere where passenger jets cruise above the level of most clouds. This prediction of a warming lower atmosphere and a cooling upper atmosphere has since been confirmed by temperature measurements from weather balloons and satellites.

It is a critical finding, partly because natural weather phenomena have less impact in the upper reaches of the atmosphere, making it easier to isolate the influence of human activity. The difference between the stratosphere and the lower atmosphere also suggests that the increase in temperature in recent decades is not due to the sun's heat increasing, as that would have warmed the whole atmosphere rather than just the part closer to the ground.

Manabe was the first scientist to create a model that produced a robust projection of the future climate, simplified to work on one of the earliest commercial computers. His breakthroughs forecast the world we are entering now, when we are halfway to doubling concentrations of CO_2 in the atmosphere.

Awarded the Nobel Prize in Physics in 2021, Manabe used his acceptance speech to draw attention to some of the consequences of the changes he had helped uncover. He focused on water, both its presence and its absence. Soil moisture would decrease in regions that are already relatively dry, he warned, while in water-rich regions, both in northern latitudes and in the tropics, there would be an increase in rainfall and the frequency of floods.

The science of a heating planet can seem intimidating, a story where the conclusions are always bleak. And yet governments refuse to listen to Cassandra's prophecy and make the changes that are necessary to deal with the threat massing outside their sea walls.

It can also feel bewildering. Humans are used to being the heroes of their story, the ones exploring and shaping the Earth, but we don't think of ourselves as having the god-like power to remake an entire planet. At the same time, changes on a planetary scale can make us feel ant-like, as if individual decisions to recycle some cardboard packaging or cut back on red meat make no difference at all in the face of the immensity of this change.

Speak to scientists and planners grappling with this, though, and they don't seem bleak. They are occasionally frustrated at not being listened to and worried for the future. But they frequently return to the fact that humanity is an adaptive species, one that's capable of harvesting far more information about the natural world than ever before and using it to make decisions.

That power was demonstrated six decades after the 1953 floods, when a severe winter storm threatened the coast of eastern England once again. As in 1953, Storm Xaver developed in the Atlantic and passed north of Scotland. But instead of sweeping down the North Sea coast, Xaver kept going east, to southern Scandinavia and into the Baltic. The 1953 storm's path meant its winds battered the east coast, but Xaver was still powerful enough to generate an exceptional surge tide.

On parts of the north Norfolk coast, the peak water levels exceeded 1953. At King's Lynn, the 2013 surge reached 6 metres. Thousands of people in coastal villages were told to evacuate. On either side of the North Sea, the Thames Barrier and the Eastern Scheldt storm surge barrier were closed. Two people were killed in the UK, but as a result of strong winds rather than the surge.

The contrast with 1953 was a vindication of the defences built in the decades since. In 2013, there were just a handful of breaches in the sea walls along a 2,800-kilometre coastline, compared with hundreds of breaches six decades before. But it was also a demonstration of the power of forecasting.

The Met Office supercomputer predicted the storm before it was even born. Drawing on satellite images, temperature, pressure and other data, it modelled the instability in the atmosphere ahead of its formation and signalled that the 2013 storm was going to track past the British Isles at the same time as a high spring tide.

A system called ensemble forecasting is now used for prediction. It works by running a computer model a number of times but tweaking the starting conditions – varying the temperature or the pressure, for example. The complete 'ensemble' of forecasts gives a span of outcomes which can be very different, especially when predicting several days into the future. An ensemble might indicate that a hurricane will range near the Gulf Coast of the US, for example, but not show precisely where it will make landfall.

To this computer modelling, a forecaster will add their knowledge of hurricanes and the global winds that guide their path in order to communicate the chances of where and how severely it will strike.

Around six days before Xaver struck, an ensemble forecast indicated a significant risk of a dangerous surge. From run to run of the modelling, there were large variations in the timing and magnitude of the forecast surge, but the warning was sufficiently clear. A few days out from the storm's arrival, there was enough confidence of the impact for the Environment Agency to give severe flood warnings.

This ability to see the future distinguishes us from the Britain of the 1950s, for whom the past was the only guide. Scientists at the Met Office are now working on computer simulations of the climate into the far future. Until recently, these simulations were run at a large scale, on grids 100 kilometres across, and at a slow speed.

Advances in computing have allowed researchers to shrink both the time and the scale, carrying out the calculations faster and going down to a grid size of 1 kilometre across, a level that is essential to capture the detail of an atmospheric process: getting down to the size of a fluffy summer cloud.

The simulation stitches together multiple aspects of the climate – temperature, wind, humidity, moisture in the soil – to give an output that forecasts a range of plausible outcomes. It's similar to the ensemble model that can warn of a storm coming next week, but peers out to the end of the century. Even now, it takes six months to produce a computer projection at 1 kilometre grid size for 20 years of the UK's future.

These plausible futures are conditional on our choices – specifically on the choice we make about the quantity of hydrocarbons we burn to power our societies. The level of global emissions used for this modelling is one of releasing large quantities of carbon with little attempt at reducing or removing the greenhouse gas.

It is not an inevitable outcome – if governments keep their promises to curb emissions – but it is a plausible one that allows everyone planning for the UK's future, from government to businesses, to identify the worst possible scenarios and prepare for them.

The most obvious consequence of living on a hotter planet is that a warmer atmosphere can hold more moisture: with every degree of warming, the water-holding capacity of the air goes up 7 per cent. The consequence is a world of more intense rain.

But there is another, contrasting, effect. Because land warms up faster than water, the relative humidity of the air above the UK decreases in future summers. This means that, in summer, there will be more intense rain showers which raise the risk of flood, but also periods of drought in between. Water companies may not be popular now, but they face an even more daunting future, one in which both draining away excess water and storing enough for dry spells will be more complex tasks.

Winter storms will increase and, as sea levels rise, storm surges will get worse, but the Met Office modelling is picking up less obvious changes too. Thunderstorms are most common in the hot days of summer, but an altering climate will bring sudden and intense downpours in the autumn too, creating an additional risk of flooding as fallen leaves block drains. The weather we experience will be a combination of the underlying shift from climate change and the natural variability of the weather, which can boost the climate effect or counter-act it.

Lizzie Kendon, head of climate projections at the Met Office, compares the effect of climate to the incoming tide on a beach.[12] 'You know that it's coming in, but the water levels don't come up gradually. They come up in a spate of waves. There is an increasing likelihood of an extreme event. That doesn't mean it will happen year on year.' The sandcastles we are building on that beach might not get knocked down immediately, but it will happen in the end.

Artificial intelligence could be used to reduce the time and cost of modelling the future. AI can be trained on existing computer models, identifying relationships within the data to develop snapshots of possible futures – a pattern of atmospheric observations in the north of Scotland that often leads to snow, for example. This would be faster as it is based on pattern recognition rather than modelling the physics of the planet. The risk is that if there are biases within existing models, the AI will learn that too.

The lesson of every past crisis that has threatened the survival of our species is that humans are capable of making rational choices and tackling immense challenges, as long as we are clear-sighted about the reality of the problem.

From surveying the Antarctic to modelling the climate, scientific endeavour has now clearly outlined the nature of the problem. The projected rise in global mean sea level means that, by the middle of this century, extremes of the sea that have been historically rare – such as the 1953 North Sea storm or the near-miss of 2013 – will become common. So common, in fact, that, by 2050, low-lying cities and small islands will experience severe coastal storms as an annual event. This will happen, the current scientific consensus suggests, under every future climate scenario that is open to us now.

After 1953, Britain and the Netherlands turned to technology to hold back floods. Countries around the planet, from Libya to Pakistan, built dams to harness the power of rivers. As populations expanded into floodplains, these attempts to master nature often had disastrous consequences. The struggle with nature is giving way to a more flexible approach. Growing concern for the environment, coupled with increasingly strained public finances, have led rich and poor countries alike to embrace more natural flood protection, while rising seas and more severe storms strain existing sea walls and barriers.

Rousseau's words about the Lisbon Earthquake in the 18th century ring true now. The death toll from floods – from the Indus Valley to New Orleans – is shaped by collective decisions about where people live and how well flood defences are maintained. There's an optimistic way to see this. If disasters can be caused by human miscalculation, making the right decisions ahead of time can save lives. We have created a world of floods. Now it's time to work out how we live with it.

ACKNOWLEDGEMENTS

Writing a book is like running a long race, the middle part of which can be a lonely trek with a seemingly infinite distance stretching ahead. But, at the beginning, there are people who are full of inspiration and encouragement.

In my case, this was Toby Mundy and Gus Brown at Aevitas Creative Management, who helped spark the idea and fan it into life. Toby is a remarkable agent, gifted with creative and commercial talent. With help from Toby and his colleague Grace Spencer, a conversation about floods turned into the outline of a book. At HarperCollins, publisher Joel Simons brought energy and enthusiasm to the project, providing clarity about its narrative drive and thematic breadth, while Erika Koljonen, commissioning editor at the Mudlark imprint, steered the book brilliantly to its conclusion. James Harding, my editor at the *Observer*, was gracious in giving me time to write, as well as letting me venture back into journalism when I missed it.

The book draws on interviews with survivors of floods, engineers, architects, weather forecasters, government planners, conservationists and academics. I am particularly grateful to Ray Howard in Canvey Island and Ria Geluk in the Netherlands, both of whom witnessed the 1953 floods from opposite sides of the North Sea. Bridget McKenzie spoke to me about her great-grandfather Herbert Abbs, while Paul Fortuin shared valuable biographical detail about his grandfather Johan van Veen.

John Curtin, former chief executive of the Environment Agency, Ivan Haigh of the University of Southampton, Alexander Hall of McMaster University and Larissa Naylor of the University of Glasgow deepened my understanding of floods and their impact on society, as did Marc Walraven in the Netherlands. Adil Najam of Boston University illuminated Pakistan's devastating floods. Laura Paul of lowernine. org provided valuable introductions in New Orleans, while Kim Heinen made many useful connections in Rotterdam.

I am grateful to Kaitlin Naughten for sharing her knowledge of Antarctica and to Jeremy Porter for his insight on the financial implications of a heating planet, while Lizzie Kendon and Grahame Madge of the Met Office helped me understand the consequences for Britain's future climate. With Wouter Helmer, I stepped back in time to explore the shifting Dutch relationship with nature. Alan Atkin-Park was my guide to the Thames Barrier.

Responsibility for this book, and the facts and storyline within it lies with me, but I am immensely grateful to every interviewee.

As a long race reaches its end, there's often a burst of energy and a chance to think about the friends and family who provided solace in the difficult moments and will join in the celebration at the finish line. I'd like to thank my friends Rob Price, Jesse Schoor, Peter Newton, Kamil Wojciechowski, Mark Maynard, Andrew Yeo and Martin Harvey, for all the beers, lunches, runs, workouts and music, or sometimes just being on the end of a phone.

My thanks above all to Meera, who gave me the time and unstinting support I needed to write. My heart is full of love for you and our children.

SOURCE NOTES

This book is based on reporting trips across Britain and the Netherlands, and on dozens of interviews that spanned the planet. It also draws on numerous written and audio sources.

I was able to interview some survivors of the 1953 floods, but many of the adult decision-makers of the mid-20th century are, of course, no longer with us. A crop of books on the 1953 floods provided valuable detail about their actions. Hilda Grieve's authoritative account *The Great Tide* and Kees Slager's *De Ramp* were particularly important. Willem van der Ham's *Johan van Veen: Meester van de Zee* shed light on that extraordinary man, as does van Veen's own book, *Dredge, Drain, Reclaim*.

The warm and enthusiastic staff at the Essex Record Office guided me through their archives, while Bert Toussaint, historian at the Rijkswaterstaat, helped me fill in gaps on the Dutch side.

PREFACE

1. 'Report of the Departmental Committee on Coastal Flooding' (1954).
2. 'Special Report on the Ocean and Cryosphere in a Changing Climate', Intergovernmental Panel on Climate Change, 2019. (https://www.ipcc.ch/srocc)
3. From the National Oceanic and Atmospheric Administration, US Department of Commerce. (https://www.ncei.noaa.gov/access/billions/dcmi.pdf)
4. Groeskamp, S. and Kjellson, J. 'The Northern European Enclosure Dam for if Climate Mitigation Fails', *Bulletin of the American Meteorological Society*, volume 101, issue 7 (2020), pp. E1174–E1189.
5. Mercer, J.H. 'West Antarctic Ice Sheet and CO2 Greenhouse Effect: A Threat of Disaster', *Nature*, volume 271 (1978), pp. 321–325. (https://www.nature.com/articles/271321a0)
6. Ed Carr, interview with the author.
7. John Curtin, interview with the author.
8. Sayers, P.B. et al. 'Third UK Climate Change Risk Assessment, Future Flood Risk: Main Report' (Committee on Climate Change, 2020). (https://www.ukclimaterisk.org/wp-content/uploads/2020/07/Future-Flooding-Main-Report-Sayers-1.pdf)

CHAPTER 1: THE SURGE

1. Bridget McKenzie, interview with the author.
2. Details of the disaster are drawn from government statements and newspaper accounts at the time, as well as reports from the 50th anniversary commemorations in 2003. McKevitt, Greg. 'Ferry Disaster Victims Remembered', BBC News, 30 January 2003. (http://news.bbc.co.uk/1/hi/northern_ireland/2705901.stm)
3. This account is based on information submitted to headquarters by the 67th Air Rescue Squadron, US Air Force, based at RAF Sculthorpe.
4. From an oral history interview held in the Essex Record Office.
5. Ray Howard, interview with the author.
6. Draws on an account published in the online archive CanveyIsland.org.
7. Grieve, Hilda. *The Great Tide: The Story of the 1953 Flood Disaster in Essex*, County Council of Essex, 1959.
8. Canvey flood survivor Rod Bishop, interview with the author.

CHAPTER 2: COUNTING THE COST

1. *The Times*, 2 February 1953.
2. Topping, Alexandra. 'Bringing a Nazi to Justice: How I Cross-Examined "Fat Boy" Göring', the *Guardian*, 20 March 2009. (https://www.theguardian.com/world/2009/mar/20/nuremberg-trials-hermann-goring#:~:text=The%20letters%20paint%20a%20surprisingly,his%20%22funny%20peculiar%22%20personality)

3. *The Times*, 9 February 1953.
4. Author interview with Rod Bishop, his son.
5. *The Times*, 21 February 1953.
6. Slager, Kees. *De Ramp*, Olympus, 2003, and articles published online by the Royal Netherlands Meteorological Institute (KNMI).

CHAPTER 3: THE LOW COUNTRY

1. Ria Geluk, interview with the author.
2. Slager, Kees. *De Ramp*, Olympus, 2003.
3. Petram, Lodewijk. *The World's First Stock Exchange*, Columbia University Press, 2014.
4. Slager, Kees. *De Ramp*, Olympus, 2003.
5. *Ibid.*
6. *Ibid.*
7. Van der Ham, Willem. *Johan van Veen: Meester van de Zee*, Boom, 2020.

CHAPTER 4: A MAN-MADE DISASTER

1. From her memo to the Waverley Committee.
2. From his report to the Waverley Committee.
3. Lord Ritchie, chairman of the Port of London Authority, in the House of Lords, 1937. (https://api.parliament.uk/historic-hansard/lords/1937/may/26/proposed-thames-barrage)
4. Callendar, G.S. 'The artificial production of carbon dioxide and its influence on temperature', *Quarterly journal of the Royal Meteorological Society*, 1938. (https://rmets.onlinelibrary.wiley.com/doi/abs/10.1002/qj.49706427503)

5. Plass, G.N. 'The carbon dioxide theory of climatic change', *Tellus*, volume 8, issue 2 (May 1956). (https://onlinelibrary. wiley.com/doi/abs/10.1111/j.2153-3490.1956.tb01206.x)

6. Author interview with Alice Bondi, Hermann Bondi's daughter.

7. Author interview with Michael Jordan, who worked with Hermann Bondi.

8. Kendrick, M.P. 'Siltation problems in relation to the Thames Barrier', *Philosophical Transactions of the Royal Society of London*, volume 272, no 1221 (1972), pp. 223–243. (https:// royalsocietypublishing.org/doi/10.1098/rsta.1972.0048)

9. While the text refers to six gates, there are in fact ten in all: six main ones that are open to shipping and four side gates. There are nine piers, numbered one to nine from north bank to south. 'Navigating the Thames Barrier', Port of London Authority. (https://pla.co.uk/navigating-thames-barrier)

10. Thatcher, M. 'Remarks visiting Thames Barrier', 21 February 1979. From the website of the Margaret Thatcher Foundation. (https://www.margaretthatcher.org/document/ 103951)

11. Memo to the prime minister by Peter Gregson, deputy secretary, Cabinet Office, February 1982. (https://discovery. nationalarchives.gov.uk/details/r/C13497477)

CHAPTER 5: SEALING OFF THE SEA

1. Author interview with Paul Fortuin, Johan van Veen's grandson.

2. Van der Ham, W. *Johan van Veen: Meester van de Zee*, Boom, 2020.

3. Van Veen, J. 'Analogy between Tides and AC Electricity'. *The Engineer*, 1947. (https://repository.tudelft.nl/record/uuid:e66ed3f0-a349-44a7-b548-ccb6785aba70)

4. *Ibid.*

5. Van der Ham, W. *Johan van Veen: Meester van de Zee*, Boom, 2020.

6. From the Rijskwaterstaat's war record. (www.rijkswaterstaat.nl/over-ons/onze-organisatie/onze-historie/rijkswaterstaat-in-de-tweede-wereldoorlog)

7. Van der Ham, W. *Johan van Veen: Meester van de Zee*, Boom, 2020.

8. *Ibid.*

9. Wemelsfelder, P. 'On the Use of Frequency Curves of Stormfloods', *Proceedings of 7th Conference on Coastal Engineering, The Hague, the Netherlands, 1960,* 29 January 1960. (https://icce-ojs-tamu.tdl.org/icce/issue/view/129)

10. Report by Professor H. Bondi of the University of London, on London flood barrier, 1966.

11. Van der Ham, W. *Johan van Veen: Meester van de Zee*, Boom, 2020.

12. Van Veen, J. *Dredge, Drain, Reclaim*, Springer, 1948.

13. 'Protest Movements Against the Closure of the Oosterschelde', *Zeeuwse Ankers*. (https://www.zeeuwseankers.nl/verhaal/protestbewegingen-tegen-afsluiting-van-de-oosterschelde)

14. Van Dam, P. 'The Amphibious Culture: A History of Coping with Disastrous Floods', 2021 lecture. (https://research.vu.nl/en/clippings/the-amphibious-culture-a-history-of-coping-with-disastrous-floods)

15. Goeller, B.F. et al. 'Protecting an Estuary from Floods – A Policy Analysis of the Oosterschelde', RAND, 1977. (https://www.rand.org/pubs/reports/R2121z1.html)

CHAPTER 6: YOU CAN'T FIGHT WATER

1. Ash, E. *The Draining of the Fens: Projectors, Popular Politics and State Building in Early Modern England*, Johns Hopkins Studies in the History of Technology, 2017; Darby, H.C. *The Draining of the Fens*, Cambridge University Press, 1956.
2. Author interview with Environment Agency official.
3. Author interview with Environment Agency official.
4. Account published on the Rijkswaterstaat website. (https://www.rijkswaterstaat.nl/water/waterbeheer/blogs-waterexperts/oof-mag-mee)
5. Kinzer, S. 'As Dike Cracks in Holland, Fear Rises with the Flood', *New York Times*, 2 February 1995. (https://www.nytimes.com/1995/02/02/world/as-dike-cracks-in-holland-fear-rises-with-the-flood.html)
6. Author interview with Wouter Helmer.
7. Author interview with Mathieu Schouten, landscape architect.

CHAPTER 7: DISASTER IN THE INDUS VALLEY

1. 'Pakistani Climate Change Minister on Floods', *Deutsche Welle*, 28 August 2022. (https://www.dw.com/en/pakistans-climate-change-minister-sherry-rehman-on-floods/video-62956190)
2. Ebrahim, Z.T. 'Climate Change is Real. So are Illegal Structures on Pakistan's Riverbeds. Will We Ever Learn?'

Dawn, 5 September 2022. (https://www.dawn.com/news/
1708388)

3. Wheeler, M. 'Harappa 1946: The Defences and Cemetery',
   Ancient India, volume 3 (1947), pp. 58–130.

4. Dales, G.F. 'Civilisation and Floods in the Indus Valley',
   Expedition, volume 7, no. 4 (1965), pp. 10–19. (https://
   www.penn.museum/sites/expedition/civilization-and-floods-
   in-the-indus-valley/#:~:text=Other%20factors%20in%20
   the%20collapse,have%20been%20a%20major%20factor)

5. Author interview with Daanish Mustafa.

6. The colonial official is Richard Temple, quoted in Gilmartin,
   D. 'Scientific Empire and Imperial Science: Colonialism and
   Irrigation Technology in the Indus Basin', *The Journal of
   Asian Studies*, volume 53, no. 4 (1994). (https://read.
   dukeupress.edu/journal-of-asian-studies/article-abstract/
   53/4/1127/338963/Scientific-Empire-and-Imperial-Science-
   Colonialism?redirectedFrom=fulltext)

7. Burton, R. *Sindh and The Races That Inhabit the Valley of
   the Indus*, London, W.H. Allen & Co., 1851.

8. Nightingale, F. 'The People of India', *The Nineteenth
   Century*, 1878. (https://scalar.lehigh.edu/kiplings/the-people-
   of-india-florence-nightingale)

9. The official is Victor Bulwer-Lytton, under-secretary of state
   for India, speaking in 1921.

10. Cunningham, J.D., *Journal of the Asiatic Society of Bengal*,
    (1849).

11. Weinraub, B. 'Pakistani Flood Refugees Stunned
    but Plan Return', *New York Times*, 1 September 1973.
    (https://www.nytimes.com/1973/09/01/archives/

pakistani-flood-refugees-stunned-but-plan-return-there-was-no.html)

12. 'A Study of Climatological Research as it Pertains to Intelligence Problems', CIA, 1974. (https://babel.hathitrust.org/cgi/pt?id=mdp.39015022217494&seq=8)

13. Supran G. and Oreskes, N. 'Addendum to Assessing ExxonMobil's climate change communications (1977–2014)', *Environmental Research Letters*, volume 15, no. 11 (2020). (https://iopscience.iop.org/article/10.1088/1748-9326/ab89d5)

14. 'Raising Temperatures: Pakistan Climate Catastrophist Sherry Rehman', *AFP*, 20 July 2022. (https://www.dawn.com/news/1700673)

15. Author interviews with Jos de Sonneville and a member of his team.

16. Khan, J. 'Centre Cuts Climate Change's Allocations by 34pc in Budget 2022–2023'. *LeadPakistan*, 15 June 2022. (https://leadpakistan.com.pk/news/centre-cuts-climate-changes-allocations-by-34pc-in-budget-2022-23/)

17. Guterres, António. 'Secretary-General's Remarks During Field Visit in Pakistan', United Nations, 12 September 2022. (https://www.un.org/sg/en/content/sg/speeches/2022-09-12/secretary-generals-remarks-during-field-visit-pakistan)

CHAPTER 8: STORM AUTOCRATS

1. 'Libya: We are your masters': Rampant crimes by the Tariq Ben Zeyad armed group. Amnesty International, December 2022. (https://www.amnesty.org/en/documents/mde19/6282/2022/en/)

2. Crawford, A. 'Frenzied, Chaotic Mess: Fears Grow Over Spread of Disease After Deadly Libya Floods', Sky News, 18 September 2023. (https://news.sky.com/story/frenzied-chaotic-mess-fears-grow-over-spread-of-disease-after-deadly-libya-floods-12964085)

3. Ashoor, A.A.R. 'Estimation of the Surface Run-Off Depth of Wadi Derna Basin by Integrating the Geographic Information Systems and Soil Conservation Service Model', *Sebha University Journal of Pure and Applied Sciences*, volume 21, no 2 (2022). (https://sebhau.edu.ly/journal/jopas/article/view/2137)

4. Gazzini, C. 'When the Dams in Libya Burst: A Natural or Preventable Disaster?', International Crisis Group, 2 October 2023. (https://www.crisisgroup.org/middle-east-north-africa/north-africa/libya/when-dams-libya-burst-natural-or-preventable-disaster)

5. Shirzaei M. et al. 'Aging Dams, Political Instability, Poor Human Decisions and Climate Change: Recipe for Human Disaster,' *npj Natural Hazards*, volume 2, (2025). (https://www.nature.com/articles/s44304-024-00056-1)

6. Malsin, J. et al. 'Libya Flood Disaster was Decades in the Making', *Wall Street Journal*, 14 September 2023. (https://www.wsj.com/world/africa/libya-flood-disasterthat-killed-thousandswas-decades-in-making-e4578d6b)

7. 'Libya: "In Seconds Everything Changed"', Amnesty International, March 2024. (https://www.amnesty.org/en/wp-content/uploads/2024/03/MDE1976082024ENGLISH.pdf)

8. Author interview with Anas el-Gomati.

9. Economist Intelligence Unit Democracy Index 2024. (https://www.eiu.com/n/campaigns/democracy-index-2024)

10. Rahman H. et al. 'Storm Autocracies: Islands as Natural Experiments', *Journal of Development Economics*, volume 159 (2022). (https://www.sciencedirect.com/science/article/abs/pii/S0304387822001249)

11. Wittfogel, K. *Oriental Despotism: A Comparative Study of Total Power*, Yale University Press, 1957.

12. AfD election programme, 2024. (https://www.afd.de/wp-content/uploads/2023/12/AfD_EW_Programm_2024.pdf)

13. Survey data shared by Steve Akehurst, Persuasion UK.

14. Author interview with Immo Fritsche.

15. Author interview with Marc Walraven.

16. Author interview with Ahmed El-Adawy.

CHAPTER 9: HURRICANE SEASON

1. Author interview with Michael Bobbitt.

2. From the National Oceanic and Atmospheric Administration, US Department of Commerce. (https://www.ncei.noaa.gov/access/billions/dcmi.pdf)

3. The senator was James Forrester, a Republican member of the North Carolina State Senate.

4. Author interview with J.T. La Bruyere.

5. These accounts are based on author interviews with Lauren Turpin, Steve White, Davida Horwitz, Angela Koh and Bryan King.

6. Morphet, J. and Nesi, C. 'Special Ops Vets Form Redneck Air Force to Ferry Aid Up Into NC Mountains After Feds

Come Up Short: Who's FEMA?' *New York Post*, 7 October 2024.

7. Edwards, C. 'Debunking Helene Response Myths', 8 October 2024. (https://edwards.house.gov/media/press-releases/debunking-helene-response-myths)

8. Gauthier, J. G. 'History and the Census: the 1803 Louisiana Purchase', United States Census Bureau, 1 April 2023. (https://www.census.gov/about/history/stories/monthly/2023/april-2023.htm)

9. State of Louisiana, Parish of Orleans, Office of Recorder of Mortgages, quoted in Seicshnaydre S. et al. 'Rigging the Real Estate Market: Segregation, Inequality and Disaster Risk', The Data Center, 2018. (https://www.datacenterresearch.org/reports_analysis/rigging-the-real-estate-market-segregation-inequality-and-disaster-risk/)

10. 'The LBJ Telephone Tapes: Inside the Presidency of Lyndon Baines Johnson', lbjtapes.org. (https://lbjtapes.org/conversation/we-need-your-help)

11. Woolley, D. and Shabman, L. 'Decision-Making Chronology for the Lake Pontchartrain & Vicinity Hurricane Protection Project', US Army Corps of Engineers, 2008. (https://levees.org/wp-content/uploads/2010/07/Woolley-Shabman-Study.pdf)

12. 'Performance Evaluation of the New Orleans and Southeast Louisiana Hurricane Protection System', US Army Corps of Engineers, 2009. (https://biotech.law.lsu.edu/katrina/ipet/Volume%20I%20FINAL%2023Jun09%20mh.pdf)

13. 'Hurricane Katrina Update', National Public Radio, 29 August 2005. (https://www.npr.org/transcripts/4822112)

14. 'A Failure of Initiative: Report of the Select Bipartisan Committee to Investigate the Preparation for and Response to Hurricane Katrina', United States Congress, 2006. (https://www.congress.gov/committee-report/109th-congress/house-report/377/1)

15. Gebauer, M. 'Will the Big Easy Become White, Rich and Republican?', *Der Spiegel*, 20 September 2005. (https://www.spiegel.de/international/new-orleans-after-katrina-will-the-big-easy-become-white-rich-and-republican-a-375496.html)

16. Author interview with Rashida Ferdinand.

17. US district court for the District of Columbia, memorandum opinion, Greater New Orleans Fair Housing Action Center et al v. United States Department of Housing and Urban Development. (https://cases.justia.com/federal/district-courts/district-of-columbia/dcdce/1:2008cv01938/133927/61/0.pdf?ts=1411522597)

18. The FEMA Katrina declaration – a letter from FEMA employees to Congress on August 25, 2025. (https://www.standupforscience.net/fema-katrina-declaration)

19. A group of scientists from the US and around the world submitted a review of the Trump administration's report as part of the public comment process. (https://sites.google.com/tamu.edu/doeresponse/home?utm_source=substack&utm_medium=email)

CHAPTER 10: MONEY

1. Author interview with Chris Hook.
2. 'Early insurance brigades', London Fire Brigade. (https://www.london-fire.gov.uk/museum/london-fire-brigade-history-and-stories/early-insurance-brigades/)
3. The Labour government that came to power in 2024 allocated £2.65 billion to 2026. (https://www.gov.uk/government/news/record-investment-to-protect-thousands-of-uk-homes-and-businesses)
4. The official is Tamara Finkelstein, permanent secretary at the Department for Food and Rural Affairs, cross-examined by the Public Accounts Committee, 27 November 2023. (https://committees.parliament.uk/oralevidence/13889/html/)
5. Batty, D. et al. 'What Lies Beneath: The Subterranean Secrets of London's Super-Rich', the *Guardian*, 7 May 2018. (https://www.theguardian.com/money/2018/may/07/going-underground-the-subterranean-secrets-of-londons-super-rich)
6. According to Doncaster council, the cost of helping 188 uninsured or underinsured properties was estimated at an average of £31,000 per property. (https://hansard.parliament.uk/commons/2020-01-30/debates/20013046000001/FloodingSouthYorkshire)
7. Shankleman, J. 'UK's Nationwide Won't Lend to Some Homes Over Flood Risk', Bloomberg, 30 April 2024. (https://www.bloomberg.com/news/articles/2024-04-30/uk-s-nationwide-pulls-mortgage-offers-to-homes-at-flood-risk)

8. Author interview with bank executive.

9. Bank of England financial stability report, November 2024. (https://www.bankofengland.co.uk/financial-stability-report/2024/november-2024)

10. Kallergis, K. 'Developer Vlad Doronin Sells Water-front Star Island Estate for Record $120m', The Real Deal, 7 March 2025. (https://therealdeal.com/miami/2025/03/07/vlad-doronin-sells-waterfront-star-island-estate-for-120m/)

11. Author interview with Jeremy Porter.

12. Top ten most significant flood events by National Flood Insurance Program payouts. (https://www.iii.org/fact-statistic/facts-statistics-flood-insurance)

13. Data from FEMA showing the cost under the previous system and the new calculation.

14. Author interview with Alys Laver.

15. Ian Liddell-Grainger, speaking in the House of Commons, 22 January 2014. (https://hansard.parliament.uk/Commons/2014-01-22/debates/140122104000001/Flooding (SomersetLevels))

CHAPTER 11: THE RISING TIDE

1. Author interview with Dave Blackwell.

2. 'Environment Agency Completes £75 million Flood Scheme in Essex', Environment Agency press release, 4 July 2025. (https://www.gov.uk/government/news/environment-agency-completes-75m-flood-scheme-in-essex)

3. Author interviews with Schoonschip residents.

4. Author interview with Cornelis Verdaas.

5. Author interview with Environment Agency official.

6. Averting Crisis: Zoning to Create Resilient Homes for All. Regional Plan Association. (https://rpa.org/news/news-release/new-report-finds-that-82-000-current-housing-units-could-be-lost-due-to-flooding-by-2040)

7. Author interview with Penny Dack.

CHAPTER 12: FUTURE WEATHER

1. 'Special Report on the Ocean and Cryosphere in a Changing Climate', Intergovernmental Panel on Climate Change, 2019. (https://www.ipcc.ch/srocc/)

2. 'Drought Reveals Ancient "Hunger Stones" in European River', Associated Press, 23 August 2018. (https://apnews.com/article/9512be71cc8f40a7b6e22bc991ef2c6c)

3. Mercer, J.H. 'West Antarctic Ice Sheet and CO2 Greenhouse Effect: A Threat of Disaster', *Nature*, volume 271 (1978), pp. 321–325.

4. The likely range under a high emissions scenario is between 0.65 and 1.1m by the end of the century. 'Special Report on the Ocean and Cryosphere in a Changing Climate', Intergovernmental Panel on Climate Change, 2019. (https://www.ipcc.ch/srocc/)

5. 'Ice Cores and Climate Change', British Antarctic Survey. (https://www.bas.ac.uk/data/our-data/publication/ice-cores-and-climate-change/)

6. 'Antarctica's Thwaites Glacier and Sea-Level Rise: Results From the International Thwaites Glacier Collaboration', thwaitesglacier.org, 2025. (https://thwaitesglacier.org/findings)

7. Author interview with Peter Davis.

8. Author interview with Kaitlin Naughten.

9. Wolovick, M.J. and Moore, J. 'Stopping the Flood: Could We Use Targeted Geoengineering to Mitigate Sea Level Rise?', *The Cryosphere*, volume 12, issue 9. 2018. (https://tc.copernicus.org/articles/12/2955/2018/)

10. 'Celebrating Lewis Fry Richardson and his Legacy', Met Office. (https://www.metoffice.gov.uk/about-us/who-we-are/our-history/celebrating-100-years-of-scientific-forecasting)

11. Manabe, S. and Wetherald, R.T. 'Thermal Equilibrium of the Atmosphere with a Given Distribution of Relative Humidity', *Journal of the Atmospheric Sciences*, 1 May 1967, pp. 241–259. (https://journals.ametsoc.org/view/journals/atsc/24/3/1520-0469_1967_024_0241_teotaw_2_0_co_2.xml)

12. Author interview with Lizzie Kendon.